COTTON
The Plant That Would Be King

COTTON
The Plant That Would Be King

BY BERTHA S. DODGE

UNIVERSITY OF TEXAS PRESS, AUSTIN

Requests for permission to reproduce material
from this work should be sent to:
Permissions
University of Texas Press
Box 7819
Austin, Texas 78712

Library of Congress Cataloging in Publication Data
Dodge, Bertha Sanford, 1902–
    Cotton, the plant that would be king.
    Bibliography: p.
    Includes index.
    1. Cotton textile industry—History. 2. Cotton
trade—History. 3. Cotton growing—History. I. Title.
HD9870.5.D63   1984   338.4'767721'09   83-23333
ISBN 0-292-76487-1

Contents

Illustrations on pp. 2, 5, 14, 28, 31, 36, 38, 42, 87, and 94 are reproduced from Edward Baines, *History of the Cotton Manufacture in Great Britain* (London, 1835); those on pp. 21, 25, 70, and 106 from Andrew Ure, *The Cotton Manufacture of Great Britain* (London, 1861); those on pp. 49 and 57 from Edna Turpin, *Cotton* (New York, 1924); that on p. 54 from Denison Olmsted, *Memoir of Eli Whitney, Esq.* (New Haven, 1846); that on p. 65 from J. G. Heck, *Encyclopedia of Illustration Sources* (New York, 1851); those on pp. 99 and 139 from James Montgomery, *The Cotton Manufacture of the United States of America Contrasted and Compared with That of Great Britain* (Glasgow, 1840).

Illustrations

Preface

It was an eighteen-month residence in Guatemala, with journeyings to remote parts of that fascinating land, that first stimulated my interest in textiles, especially those woven from native-grown and -spun cottons, leading to a personal collection of nearly two hundred distinct ones. Since then, visits to Peru and Chile have further increased my interest. For the native textiles of those lands are more than body coverings; they are an art form that is socially significant. In Guatemala especially, textile design and weaving techniques are so much a part of the heritage of each Indian town that even a skilled weaver hesitates to copy design and techniques from another town.

Once thus stimulated to consider textiles as more than body coverings, I became interested in the history of all textiles, then inevitably in the impact of textile fibers, especially plant fibers, on the world's history.

So this book was born. It does not pretend to analyze in minutely documented detail every step of the progress from those "green withes" of Samson's day to present-day synthetic fibers, but it tries to give an overview of such progress and relies heavily on a great number of previous writers whose works are listed in the bibliography. For easy access to such books as I do not own personally and for the occasional procuring of a book by interlibrary loan, I must thank the many patient librarians of the Bailey-Howe Library of the University of Vermont who have helped me on my way. Botanical details have been checked with a professional botanist, my husband Carroll W. Dodge, though he is not to be blamed for any errors, stated or implied, that I may have made.

I dare hope that some readers may become as enthusiastic about textiles and their role in history as I have.   B.S.D.

COTTON
The Plant That Would Be King

1. The Enthronement

When, on March 4, 1859, the Honorable James Henry Hammond rose to address his colleagues in the United States Senate, he had every reason for supreme self-confidence, or so he, as well as many of his colleagues in the Senate, believed. An editor of *The Southern Times*, he knew himself to be as skilled with words as with the politics that had previously placed him in the United States Congress, then in the governor's seat of his native state, South Carolina. But what really gave him status and justified him in speaking for fellow Southerners was the plant that covered the large estate he had acquired some time before along with a bride of ample fortune.

A shrubby plant with deeply divided leaves and with flowers somewhat resembling those of the hollyhock, it thrived in the hot and humid atmosphere of South Carolina and the neighboring states and produced a golf-ball-sized fruit which splits on maturity to reveal a mass of dark seeds embedded in soft down. It was this down, this almost impalpable bit of fluff multiplied by millions, that had made the fortunes of such as the Honorable James Henry Hammond. It was the future of this fluff that had long been obsessing members of the United States Senate during the months before Senator Hammond made his speech. The Southern contingent was struggling to modify the Missouri Compromise of 1820 and thereby pave the way for elevating Kansas Territory into statehood dedicated to that peculiar institution, slavery, upon which the South's profitable cotton crops depended. Should they fail in this device for keeping slavery and cotton raising intact, the states where both had long flourished were prepared to leave the Union and, should that move be challenged, to venture life, fortune, and sacred honor to maintain their stand.

1

The Enthronement

"Would any sane nation make war on cotton?" Senator Hammond demanded of his colleagues, then answered the rhetorical question himself. "Without firing a gun, without drawing a sword, should they make war on us, we could bring the whole world to our feet. The South is perfectly competent to go on one, two, or three years without planting a seed of cotton. . . . What would happen if no cotton were furnished for three years? . . . this is certain: England would topple headlong and carry the whole civilized world with her, save the South. No, you dare not make war on cotton. No power on earth dares make war on cotton. Cotton is King."

To be sure, cotton and the processing of it had led England, by way of its burgeoning textile factories, into the Industrial Revolution—a process which, Senator Hammond believed, could not be reversed without national disaster such as only cotton shipments from his South could prevent. Nevertheless, beyond such dire forecasts of evils to come, the senator seemed to concern himself not at all with the role cotton had already played in history and very little with its influence in areas outside of that commerce upon which his own and his fellow planters' well-being rested. Inquiries as to the antecedents and future of royal personages seemed not quite proper. The king was there to accept homage due and, in turn, to reward the loyalty of his subjects. At the time of Senator Hammond's speech, nothing else seemed to matter. For most of his colleagues there was to come a time of reckoning, but, as it was to turn out, the senator would not be alive to draw a lesson from it.

That lesson could be summed up in a few sentences—the plants that, for various reasons, people let themselves become dependent upon come to exercise a tyranny out of all proportion to their own intrinsic worth. Thus it had been with the Far Eastern spices which early European navigators brought home at a cost, as one historian estimated it, of one thousand lives per cargo. Thus it had been with a Brazilian jungle tree whose coagulated sap served natives for waterproof mantles and for playthings like bouncing rubber balls and squirt guns. And thus it was, from the start, destined to be with that harmless bit of fluff called cotton.

2. Fiber into Fabric

That bit of fluff—so fine that a breath of air could blow it away, so powerful that it could change the destiny of nations, challenging people to invent machines, then to invent ways to power those machines, and finally to produce fibers without the intervention of seed and plant—could never have come to preside over a kingdom had not spinning been first invented. Spinning has always had a hold on the imagination of people who find a special fascination in the spinner's power to transform fine fibers into long, strong threads that may further be spun into ropes or set on looms to be woven into fabrics of infinite variety and design.

On many a frieze carved in ancient Greece or Rome sit those three sister Fates in whose hands rests human destiny: Clotho, the youngest, wielding a distaff from which she spins the thread of life; Lachesis, the middle one, measuring off that thread for the inexorable Atropos to sever. In later times, when the pagan legends of ancient Greece and Rome had become outmoded, the magic motif survived. Many children of our own times have been entertained by centuries-old folktales in which some much abused maiden, being set the task of spinning a roomful of flax into gold, succeeds with the help of some superhuman being, to end in the arms of a Prince Charming whose throne she will eventually share.

It was not the flax but the spinning of it that really counted, and spinning could be performed with other fibers like wool or cotton, though the latter was a latecomer on the European scene. Once processed, any such fibers could produce income and power for the processor, incidentally revolutionizing the lives of nations and assuming the role of unelected tyrant.

Gold it was, though never the kind of gold to light fires of covet-

4

ousness in the eyes of men like the Spanish conquistadores, who were to encounter not only the cotton plant but the spinning and weaving of it among the heathen natives of the lands they were claiming for Spain and for the Holy Catholic Church. The plant and the fiber could have been no novelty for them, since it must have reached Spain out of the Near East, along with the Moors who overran the Iberian Peninsula during the eighth century. A frail and perishable commodity without the luster or exchange value of gold or jewels, it did not even stir the swashbuckling beholders to describe the woven cottons they encountered in the New World or to tell of the way they must have seen the fiber being spun and woven.

To today's beholder, this conversion of a handful of fluff into usable threads still smacks of magic, even where the device used to achieve this feat is no more sophisticated than half of a gourd shell, in which is twirled a foot-long slender stick bearing a sticky ball of clay about four inches from one point. You can see spinners at work on almost any market day in some highland Guatemalan Indian town. There, in the central plaza, looked down upon by gleamingly whitewashed government buildings and an equally gleaming Spanish colonial church, Indian women have forgathered to make a bit of profit if possible and, in any case, to enjoy the social occasion of market day.

Wearing a huipil, the exquisitely designed and woven blouse characteristic of her own town, and a hand-loomed cotton skirt decorated with intricately tie-dyed patterns, the spinner squats on the hard-packed dirt of the plaza. Before her is a circle of little baskets or half-gourds containing the produce—squash or beans or corn or avocados—which explain and justify her presence there. While she and her almost identically clad neighbors keep up a stream of chatter, their hands are busily picking seeds out of bolls of cotton, white or native brown, which she has brought in the half-gourd resting on the ground beside her. Once she has a gourd full of clean cotton and has pulled each boll out into a soft mat, she is ready to start spinning.

The spinner picks up her spindle—that slender pointed stick—and rests one end in the smaller half-gourd. She presses onto the clay ball, which is now nearer the lower end of the stick, the end of one mat of cotton. Then she starts twirling the stick with her right hand while her left hand, above and to one side of the twirling stick, holds a handful of cleaned cotton. At the upper tip of the twirling stick the impalpable bit of fluff becomes a firm thread. The Indian woman will pause to wind that thread upon the spindle's clay ball before pressing more fluff onto that which remains in her hand. During all this, her chatter never misses a beat, for to her and to her cronies it's all a routine which must have been old at the time the conquistadores arrived on the scene. For today's visitor from far and alien lands, it still has an aura of magic.

That doughty old conquistador, Bernal Diaz del Castillo, who spent the last years of his adventurous life in Antigua Guatemala recording his memories of the conquest, never gives us a picture of Indian women at this work, which must then, as today, have been a commonplace in the market of Antigua Guatemala. But he does mention, though never describes, the *mantas* of cotton which were sent to Montezuma as tribute. Even the Spaniards must have considered these *mantas* to have had some value, since they were soon transferring that kind of tribute to their own monarch.

It was a barefoot Franciscan friar who arrived in Mexico in 1523 and thereafter traversed Mexico from one town to another, converting Indians by the thousands and keeping sporadic records of what he observed, who described textiles such as those which, on Easter day, 1536, the Christianized Indians of Tlaxcala gave as offerings to the church.

Fray Toribio de Benavente, called Motilinia by the Indians, wrote of *mantas* "woven of cotton and hare's wool, and those are many and of many kinds. Most of them have a cross in the middle, and these crosses differ much among themselves; other such cloths have in the middle a shield showing the five wounds [of Christ] woven in colors; others have the name of Jesus or Mary, with tassels and embroidery all around them, and this year a woman offered one such cloth with the crucifix woven on both sides, although one seemed to be the face of it, and this was so well done that all those who saw it, both churchmen and lay Spaniards, admired it greatly" (pp. 83–84; my translation).

Professional weavers still greatly admire that *manta*'s lineal descendants, woven today in towns like San Antonio Aguascalientes near Antigua, Guatemala. The so-called *tzut*, which serves as anything from basket cover to a carrying sling for a baby and which is less formal than the blouse or *huipil*, exhibits any kind of woven pattern the weaver's fancy may dictate. It is so fashioned that the design on the reverse is practically identical with the one on the front—something that few weavers in more sophisticated lands can hope to duplicate. Other towns show differing designs and techniques, employing any kinds of colored threads the weaver can purchase.

Wool, in our day, comes from sheep that are the lineal descendants of those long ago introduced from Spain. Silk may now be the genuine article rather than that ancient type described by the historian Prescott as from "a species of caterpillar, which spun a thread which was sold in the markets of ancient Mexico." But cotton is a native American plant, and it was through the spinning of it that native American Indians long ago came to evolve their incredibly beautiful and intricately woven textiles.

Garcilaso de la Vega, the mestizo son of a conquistador of Peru and an Incan princess, gives us a picture, though a tantalizingly brief one, of the Peruvian Indian women and their spinning. Having taken up residence in Spain, Garcilaso undertook to interpret the land of his own birth for the people of the land of his father's birth. Of the women of Peru, he wrote, "They busied themselves with spinning and weaving wool in the cold districts and cotton in the hot. Each woman spun and wove for herself and for her husband and children." Then he added a few sentences that underline the kinship of the Peruvian women and their craft with the Guatemalan: "Every piece of cloth they made, for whatever purpose, was made with four selvages. Cloth was never woven longer than what was needed for a single blanket or tunic. Each garment was not cut, but made in one piece, as the cloth came from the loom, and before weaving it they fixed the approximate length and breadth." It's hard to imagine cloth woven to measure, but that was clearly being done four centuries ago in Peru. It's also what the Indian women of Guatemala have done in our own century—looping warp threads over cords that run close along the sticks of their stick looms, then weaving right up to the ends of those loops. These weavers who had to spin their own threads figured out all possible ways to save that precious thread.

Spinning in sixteenth-century Peru more resembled that of ancient Rome than the present-day Guatemalan. Ancient Roman bas-reliefs show both distaff and spindle. The distaff was a stick two or three feet long, and sometimes highly ornamented; it was thrust through a soft ball of wool or flax fibers (cotton then being unknown in Europe) so rolled that the fibers could be drawn out. Holding the distaff under her left arm, the spinner, with the fingers of her right hand, drew out the fiber and gave a quick twist to the spindle, to which the thread was attached, as she let it drop to the ground. This spindle, about one third as long as the distaff, had a small disc set at the lower end to steady it and a slit at the top through which the thread was passed after it had been spun and wound on the spindle, as Guatemalan women wind their threads on their twirling sticks. In ancient Italy, it was considered an ill omen for a woman to spin while walking.

Peruvian women were either more skilled or less superstitious. As Garcilaso told it:

The Indian women were so fond of spinning and so reluctant to waste even a short time that as they came or went from the villages to the city [of Cuzco] or even from one quarter to another . . . they carried equipment for the two operations of spinning and twisting [into heavier yarns]. As they walked along, they twisted what they had spun, this being an easier task. While visiting, they would take their distaff and spin as they conversed . . . The spindles were of cane . . . They cast a loop around the spindle from the thread they are spinning, and as they spin, they drop the spindle as they do when they twist. The thread is made as long as possible. They pick it up by the middle finger of the left hand and pass it on to the spindle. The distaff is held in the left hand . . . it is a quarter of a vara [about seven and one-half inches] long. They hold it with the two smallest fingers, and use both hands to thin the thread and smooth out the burls . . . In my time they did not spin flax, which was unknown, but only wool and cotton.

His only agricultural note in regard to cotton seems to be in this sentence: "On the llanos, or seacoast, where the climate is hot . . . they make cotton cloth from cotton grown on the land of the Sun and of the Inca." This observation tells us nothing about the early Peruvian method of planting and harvesting cotton, but it does suggest that cotton, to have been grown on lands reserved for the sun and the Inca, must have been considered a precious commodity.

It was inevitable that a social ritual became involved in so important and so universal an occupation. Garcilaso gives an amusing sketch of how the ritual worked. When a woman of a lower social class presumed to pay a social call upon a woman of the upper class, she would show her awareness of the social gulf by suggesting that her hostess give her some spinning to do. The hostess would then graciously put her visitor at ease by offering to let her share in some task in which she or her daughters might be engaged. Clearly the spinning and weaving of wool or cotton claimed a very important part in the lives of all Peruvian women.

Wool garments, as Garcilaso pointed out, could be uncomfortably hot, and flax was unknown in Peru. Having told of the homesick Spanish lady who had sent home for flax fiber and a loom, he wrote "as this [1560] was the year in which I left Peru, I did not know whether she received them or not. I have since heard that a great deal of flax has been gathered, but I do not know how well my Spanish and Mestizo relatives have turned out as spinstresses, for in my time, as they had not flax, I never saw them spinning it . . . They had, of course, fine cotton and excellent wool which the Indian women spun marvellously. They used to card these between their fingers, for the Indians never invented teasels and the Indian women had no spinning wheels." In actual fact, ancient Peru had no wheels of any kind, for they did not provide a very satisfactory means of traversing the steep, rocky trails of that high and mountainous land. And without wheels, the spinning and weaving of cotton must have remained a very limited, very personal craft.

3. The Fabric Finds a Future

Cotton is a multinational plant, and one or another of its many varieties is native to most of the warmer regions of this earth. Long used by human beings, it was, nevertheless, slow to acquire snob value. No yearning for fine cottons sent sixteenth-century galleons roaming distant seas. No greed for controlling lands where cotton grew native set nations of the seventeenth century at odds. In fact, though older than recorded history and native to romantically distant lands, cotton was to make no great stir with Europeans until after the nineteenth century had dawned and the age of machines was ushered in. No one could then have been persuaded that this shrub, with leaves much like the grape's and a fruit that bursts open to reveal seeds embedded in a soft boll, was destined to play a large and dramatic role in the affairs of nations.

Pliny, the Roman historian of the first century of our era, wrote of the plant which he considered an exotic curiosity: "the upper part of Egypt, lying in the direction of Arabia, grows a bush which some people call cotton. . . . It is a small shrub, and from it hangs a fruit resembling a bearded nut, with an inner silky fiber from the down of which thread is spun. No kinds of thread are more brilliantly white or make a smoother fabric than this. Garments made of it are very popular with the priests of Egypt" (vol. 5, p. 429).

Cooler, more comfortable and more easily washable, less tantalizing to fiber-consuming insects than woolen garments, and made from a fiber more easily prepared and spun than flax, brilliantly white cotton robes were bound to appeal to priests, who may have decided to keep such rare and wonderful materials out of the hands of ordinary Egyptians. If in those times any such fabrics reached the European shores of the Mediterranean, they must have been regarded as exotic and desirable. Yet they were not so desirable as to cause even an imperial-minded Rome to compete for control of far-off lands that produced a plant whose fibers few Romans knew anything about and fewer still could have come to value. Even Pliny was accepting that cotton fabrics belonged to a distant, exotic land.

Fifteen centuries were to pass before Europeans began to take an active interest in the plant itself and then only as one among the many botanical curiosities featured in the newly popular herbals of the day. John Gerard's *Herball* gives it due, but not over-long notice, as "The Bombaste or Cotton Plant." He describes it partly from what he had been able to grow from seed in his own botanic garden and partly from accounts given him by English travelers to the Levant, where the growing season was long enough to permit it to mature. Gerard wrote:

> The Cotton bush is a low and base plant, having small stalkes of a cubit high and sometimes higher, divided from the lowest part to the top into sundry small branches, whereupon are set confusedly or without order a few broad leaves, cut for the most part into three sections, and sometimes more, as Nature list to bestow, somewhat indented about the edges, not unlike to the leafe of the Vine . . . but lesser, softer, and of a grayish colour: among which come forth the floures, standing upon slender foot-stalkes, the brimmes or edges whereof are of a yellow colour, the middle part purple; after which appeareth the fruit, round and of the bignesse of a Tennise ball, wherein is thrust together a great quantitie of fine white Cotton wooll; among which is wrapped up blacke seed of the bignesse of peasen, in shape like the trettles or dung of a cony. The fruit being come to maturitie or ripenesse, the husk or cod opens it selfe into foure parts or divisions, and casteth forth his seed and wooll upon the ground, if it be not gathered in time and season. The root is small and single, with a few threds anexed thereto, and of a woody substance, as is the rest of the plant.
>
> It groweth in India, in Arabia, Egypt, and in certaine Islands of the Mediterranean sea, as Cyprus, Candy, Malta, Sicilia, and in other provinces of the continent adjacent. It groweth about Tripolis and Aleppo in Syria, from whence the Factor of a worshipfull merchant in London, Master *Nicholas Lete* before remembered, did send unto his said master divers pounds weight

of the seed, whereof some were committed to the earth at the impression hereof, the successe we leave to the Lord. Notwithstanding my selfe 3 yeares past did sow of the seed, which did grow verie frankly, but perished before it came to perfection, by reason of the cold frost that overtook it in the time of flouring. (pp. 900–901)

Those were the days when an ambitious herbalist planted every seed that came his way in a "physic garden." He was primarily a pharmacist, engaged in the unceasing search for cures of ills whose causes very few people then understood. Nature, they believed, suggested the possible curative uses of plants in the plants themselves. For most of the plants Gerard described, there is a list of such "vertues" almost as long as the description of the plant itself.

Though the cotton plants Gerard attempted to raise did not mature to produce viable seed, his plants would have developed sufficiently to show they could produce quantities of seed. Hence, he hesitated not at all in ascribing to the plant aphrodisiac powers: "It stirreth up the lust of the body by increasing naturall seed, wherefore it surpasseth." One has to wonder on whose testimony he was relying. And how was that seed administered? Not through its pressed-out oil, since he ascribed to this oil quite distinct "vertues": "it taketh away many freckles, spots, and other blemishes of the skin."

Finally came the part of the plant that was to be remembered when all such pseudopharmaceutical uses were forgotten: "To speake of the commoditie of the wooll of this plant were superfluous, common experience and the dayly use and benefit we receive by it show them. So it were impertinent [i.e., not pertinent] to our historie to speake of the making of Fustian, Bombazies, and many other things that are made of the wooll thereof." Perhaps that implied "impertinence" was a consequence of the herbalist's limited acquaintance with spinning and weaving. Had he ever watched spinsters at work? Had no one told him that cotton thread, as then produced, lacked the strength to serve as warps, which were either linen or wool, resulting in those hybrid fabrics called "Fustians" and "Bombazies"?

However that may be, no sixteenth-century herbalist could have realized that cotton, since assigned to the genus *Gossypium*, boasted many fiber-producing species. Some of these, as the *Gossypium* of India, grow into trees that may reach a height of ten or more feet. Other species, growing in India or China, produce a white, reddish, or tan lint with a dense reddish fuzz attached to the seeds. The American species, which were to modify national, international, as well as textile history, were the common Upland cotton and the much sought-after Sea Island cotton whose ancestors may have been immigrants from the Barbados. Also American are species native to Central and South America, varieties of which may be a golden tan or perhaps also greenish or bluish, as sometimes reported. The genus is large, with many centuries of cross-breeding to produce strains that have confused some experts in the past.

Among the careful observations of the Far East were those of the eighteenth-century Dutch sea captain and later rear admiral Jan Splinter Stavorinus:

> Cotton is likewise a product of Java. The shrub that produces it is cultivated in almost every part of the island by the natives. . . . The Dutch East India Company, to whom the greatest part of it is delivered, pay for it, according to its qualities. . . . The largest part of the cotton-yarn produced is sent to Holland; the rest is employed by the natives, in weaving cloths for their own consumption. . . . Attempts have likewise been made to introduce the manufacture of cotton cloths, as an article of trade for the Company, and to supersede part of their large importations of the article from *Hindostan.* (vol. 3, p. 330)

Of fabrics woven in Bengal, Stavorinus notes: "The chief articles of commerce which the country yields are silk, muslins, calicoes, cottons and other piecegoods. . . . The materials from which their piecegoods are wove is the *capok*, or cotton." This "capok" is not to be confused with today's unspinnable kapok, which grows on a very tall tree. Stavorinus continues: "It grows upon a shrub, or tree, which is cultivated in very great abundance, in this country, though yet not in sufficient quantities for all the piecegoods which are annually manufactured here; for much of it is brought hither from Surat" [on the opposite side of the Indian subcontinent]. "Some kinds of piecegoods, likewise, require two different sorts of *capok*" (vol. 1, pp. 471, 473–474). The latter is a tantalizing comment, since the "different sorts" are not described.

According to Stavorinus, "The *capok* is stretched with a wire, upon an arched piece of wood, like a bow, cleaned from all its impurities, spun by the women into yarn, and finally woven into piece goods of various denominations by the men. The weaving manufactories are dispersed throughout the country . . . a distinct kind is wove in every district." Those various denominations of piecegoods included the printed cottons, commonly called chintzes, and the "finest muslins and cottons."

Inevitably, employees of the English East India Company had been charmed by the cottons which, especially in a hot climate, were so kind to the skin. They were soon sending some home to England, undoubtedly as gifts but also as a form of remitting funds. The home folks were equally delighted with the new, light, soft, cool fabrics from the Orient with their colorfully printed designs. In fact, their delight was so potent, with the new fabric becoming so fashionable, that imported cottons were soon being seen as a menace to the place long occupied by traditional English woolens.

This was not a situation to be meekly endured by a Britain where sheep raising, wool spinning, and weaving had long involved the interest, the activities, the pocketbooks, and hence the lives of so many Britons. By 1700, thoroughly alarmed and undoubtedly under pressures from their constituents, members of Parliament had decided to put a stop to it all by passing a law limiting the importation of cotton fabrics. It did not seem necessary to include raw cotton fiber in the interdiction—a loophole on which cotton fanciers soon zeroed in. Presently, the manufacture of cotton piecegoods was to be almost entirely transferred from India to Britain.

That England had on hand many skilled weavers eager to accept the challenge of a new kind of fiber must be credited, in part at least, to a man who would have been delighted had England and all her heretics disappeared into the sea—Philip II of Spain, who conceived it to be his mission to eliminate all heresy from the face of the earth. On August 17, 1585, when the great Flemish commercial city of Antwerp capitulated to Philip's cousin, the Duke of Parma, military commander in Flanders, Philip made it clear that the heretics must choose between renouncing their religion or going into perpetual exile. Most chose the latter, as shown by a letter the Duke of Parma sent from Antwerp to his cousin in November of the same year: "the poor city is most forlorn and poverty-stricken, the heretics having all left" (quoted by Motley, vol. 1, p. 261).

Having chosen what was to them the lesser of the two great evils, they had then to choose where they should spend their lives of exile. It must, of course, be a safely Protestant land—possibly Lutheran Germany or Calvinistic Switzerland. Or possibly England, ruled by a heretic queen whose power Philip was to challenge three years later in the disaster of the Spanish Armada. A historian of the English textile industry, discussing its origins, mentions the belief that it may have been introduced from Turkey:

. . . but I am more inclined to think that the art was imported from Flanders . . . by the crowd of Protestant artisans and workmen who fled from Antwerp, on the capture of that great trading city by the Duke of Parma in 1585; and also from other cities of the Spanish Netherlands. Great numbers of these victims of sanguinary persecution took refuge in England, and some of them settled in Manchester; and there is stronger reason to suppose that the manufacture of cotton would then be commenced here, as there were restrictions and burdens on foreigners setting up business as masters in England, in the trades then carried on in the country, whilst foreigners commencing a *new* art would be exempt from those restrictions. The warden and fellows of Manchester college had the wisdom to encourage the settlement of the foreign clothiers in the town, by allowing them to cut firing from their extensive woods, as well as to take the timber necessary for the construction of their looms, on paying the small sum of four-pence a year. (Baines, p. 99)

Whether enough raw cotton fiber was being imported into England in the last decade of the sixteenth century to sustain the exiled Flemish woolen weavers in a cotton trade must seem a bit doubtful. Nevertheless, as skilled wool spinners and weavers following a family trade from one generation to another, they would have found some employment in the craft, and their descendants would have been on hand in the early eighteenth century to take part in the new industry, which was then in the process of being almost totally transferred from India to England. By 1719 this industry had grown to such proportions that the woolen interests, through Parliament, again tried to have the importation of raw cotton interdicted, this time meeting with no success whatever.

Raw cotton fiber continued to enter Britain, and the new breed of cotton manufacturers continued to grow in wealth, power, and world influence throughout the succeeding century. Manchester and nearby Lancashire were on their way to becoming busy centers of cotton weaving as well as spinning, voraciously consuming shipments of fiber from India, Egypt, and the Levant. Similarly the West Indies and the American colonies were shifting their allegiance from King George to King Cotton. The export of cotton from the South, where the growing season was long enough to permit it to mature, advanced by leaps and bounds until it began to play a crucial role in international relations. In 1791, the United States was producing two million pounds of cotton fiber. By 1860, that figure had leaped to over sixteen hundred million, with by far the largest percentage going to England.

Two years after those 1860 shipments, the only cotton reaching England from the United States was contraband. This dramatic drop in cotton exports and imports not only shook King Cotton on his throne but threatened to topple the throne itself. Most of all, it reached out into the lives of King Cotton's subjects—the planters and their slaves, the factory owners on both sides of the Atlantic, the factory workers left empty-handed in empty mills. Beyond that, it played a large role in the destinies of other lands whose cotton crops, it was hoped, might fill the supply gap. All in all, the multinational plant was becoming an incalculably large factor on the international scene.

The Fabric Finds a Future

For all this, cotton was really not to blame. It remained a harmless bit of fluff. Nor could any one person or group be blamed. As long as cotton fabrics were a bush-to-back proposition, as in Guatemala, or as long as their production remained a cottage industry, cotton had little power to alter anyone's life. The menace arose when cotton processing was moved into factories, whether they were dirty and wasteful or clean and efficient. For they all had one important thing in common—all had come to demand huge quantities of raw material grown oceans away, from a shrub that usually was planted every year, and harvested by hand, and from a fiber then picked clean, baled, and shipped.

It was this soaring demand and the ways of coping with it that were to make cotton's role crucial in the affairs of nations. It's an infinitely complicated story and a fascinating one, if only one can summon the patience to study the business from the start, to unravel the intricate web as one might an ancient and delicate fabric—with patience and care, to discover just how it was woven by people who could not often be troubled to leave records of what they did and why they did it.

4. Looms Speed Up

Once a Guatemalan Indian woman had made herself and her family all the garments they needed, adding perhaps a *tzut* to cover either her head or her market basket or perhaps to use as a back pack for her baby, she might make an extra garment or *tzut* to sell to a fellow townswoman or to an admiring tourist whom luck might bring her way. Her idea of profit would seem as primitive as her methods of converting fiber into yarn, for she would have no idea of counting the hours she spent in the work and reckoning her *tzut* at so much per hour. The exquisite weavings that have found their way into the textile museums of many lands would have been sold at a simple cost-plus price—the cost of materials such as the purchased threads that enhance the design, plus whatever she had the courage to ask above that figure.

What, in her simplicity, the Guatemalan Indian dared demand could certainly never account for the most minimal wage for the un-counted hours of spinning and weaving that go into even the sim-plest four-selvaged *tzut*. Nor could there be much progress, as the outside world reckons it, where each family takes responsibility for its own necessities and time counts for so little that no one bothers to devise ways to simplify labor.

Time itself may not have been of much greater value to the Indians of the Far East, but textiles were. Since the beginning of history, the textiles of India, notably the cotton ones, were an article of trade in Far Eastern lands. And once the ships of Europe had found ways to penetrate Far Eastern seas, textiles from the Orient would find their

Note: Unless otherwise indicated, all quotations in this chapter are from Edward Baines, *History of the Cotton Manufacture in Great Britain*, Chapters 8–10.

way back to Europe in quantities immeasurably greater than the almost negligible amounts that, like the earliest spice shipments, had had to come by caravan over perilous, brigand-infested routes. Rare, costly, and coveted, cottons were soon in high fashion and high demand.

Wool and flax had long since taught European workers a great deal about textiles in general—how fiber should be prepared for spinning, how the yarns should be spun, and how the spun yarns should be used in weaving textiles. Thus, as demand for cotton soared to the point of competing heavily with the local woolen industry and as protection for that industry was consequently being sought, English folk who saw themselves being deprived of the cottons they so liked decided to do something about it. Why shouldn't the English themselves spin and weave cottons, as they had been spinning and weaving woolens and linens?

To do this, they had to have quantities of raw cotton within reach. This access was no simple matter when all cotton grew continents away. Once they had it in hand, they had to learn how to spin a fiber totally different from the wool or flax they had known. For professional spinners, this would have seemed relatively easy. No one, of course, could see any kind of troubles ahead. Nor would there be any of a serious kind as long as spinning and weaving remained cottage industries. They had a simple trust that there was bound to be enough such fiber in the world and that someone would locate it and see to it that it reached the doors of English cottages, whether it came from India, Egypt, or America.

For long no one seemed concerned over the fact that there were many steps from seed to baled cotton fiber and that should a single step lag, all others must wait upon it. If enough seed had not been planted, if the weather was hostile, if the soil had not been of the right quality, or if the plants had been improperly cared for and thus at the mercy of insect and fungal diseases, cotton pickers would not be able to bring in the large crop expected and those slaves assigned to pick out seeds and bits of waste would not be kept occupied. If, on the other hand, there was a very big crop, there might not be enough hands to cope with the picking and cleaning. In short, if somewhere along the line, from seed to finished fabric, supply did not equal demand, people must wait idly by and profits be limited. And, as always, people counted most.

Weaving, of course, is always what spinning is about, for a finely spun thread by itself can amount to little until it becomes a rope or part of a finished fabric. This conversion of thread into fabric demanded new skills determined by the particular kind of textile to be woven. But all textiles, however intricately woven, owed their existence to one or another kind of loom. In Guatemala it may look like a bundle of sticks, but with this bundle a knowledgeable weaver can create a blouse fabric that is a marvel of intricate beauty. For the wider skirt materials, generally woven by Indian men, Guatemala has looms that do not differ too greatly from the cottage looms of any other land, save that native hardwood cogs may replace imported metal.

A cottage loom such as the English or American weavers of the eighteenth century used looked something like the frame of a huge packing box. Massive wooden timbers were needed to withstand the inevitable constant vibrations of a loom in use. This frame has two wooden drums, the warp beam and the cloth beam, and is provided with a cogged mechanism to hold the threads between them taut. In the space between the two rollers, the warp threads are drawn through the eyes of two or more harnesses, alternately where they number only two. Thus each harness could be raised in turn to create between the separated threads a "shed" through which a shuttle bearing the weft thread might be passed. Then the positions of the two harnesses are interchanged and the new weft beaten up with a horizontal bar which, on the large cottage loom, swings freely from an attachment above. Again the shuttle is passed through the shed and the whole process repeated indefinitely.

The harnesses of a Guatemalan stick loom are loops of cord hanging from wooden bars. The beating up is done by a long, flat, highly polished strip of hardwood. However, the principle of this loom does not differ from that of the cottage loom. On both, weaving is accomplished by alternating the positions of the harnesses, beating up the weft, throwing the shuttle. Should a fabric of more complicated weave than homespun be desired, more harnesses can be used and the sequence of raising and lowering adjusted to the required fabric. Today's power looms are basically no different from the old cottage looms, except that every step is done through prior automatic settings.

Even in a cottage industry, there were bottlenecks once the demand for cotton fabrics had started to grow. Yet had efficiency reigned all along the line, a weaver might still have found it hard to keep the loom busy. Strung with warp threads of linen (long cotton threads were not strong enough for warping), those hybrid fabrics called "fustians" usually demanded for a single loom more cotton weft than could be spun by all the womenfolk of a weaver's family together. It has been variously estimated that anywhere from four to eight spinners were needed to keep a single loom supplied with thread. So, to fill a quota of cloth, a weaver had to enlist the skills of outside spinners.

Of the eighteenth-century English cotton weavers, Baines wrote: "It was no uncommon thing for a weaver to walk three or four miles in a morning and call on five or six spinners, before he could collect weft to serve him for the remainder of the day and when he wished to weave a piece in a shorter time than usual, a new ribbon or gown was necessary, to quicken the exercises of the spinner."

The urgency to maintain the supply of spun threads for the looms that produced fabrics to clothe families of early colonial days is evinced by those interesting survivals, the rockees. The spinner sat at one end of the rockee, working the treadle of the spinning wheel; this motion gently rocked to sleep a baby placed in a fenced-in section at the other end of the rockee.

Perhaps the general discrepancies of supply and demand might somehow have been adjusted had not a new factor entered the equation—invention, fueled by demand and fueling further demand as the industry grew apace and the new breed of industrialists thus being created had their appetites for profits whetted. Profits, it was soon realized, could be made to grow faster if only some way could be found to shorten the time from raw fiber to finished fabric. Such an invention was to affect the lives of millions and the relations of nations.

Logically, the first steps in this direction should have begun with speeding the production of the fiber itself, following that with mechanical spinning devices. A final development should have been the mechanization of looms. Illogically, it started where it should have ended, by making weaving easier and faster—a foolish step, one might think, in view of the fact that the demand for spun yarns already so exceeded the available supply that, as Baines noted, "the seller could put her own price."

The seller probably enjoyed this position of power, creating a difficulty which

> was likely to be further aggravated by an invention which facilitated the process of weaving. In the year 1738, Mr. John Kay, a native of Bury, in Lancashire . . . suggested a mode of throwing the shuttle, which enabled the weaver to make nearly twice as much cloth as could be made before. The old mode was to throw the shuttle with the hand, which required a constant extension of the hands to each side of the warp. By the new plan, the lathe (in which the shuttle runs) was lengthened a foot at either end and by means of two strings attached to the opposite ends of the lathe, and both held by a peg in the weaver's hand, he, with a slight and sudden pluck, was able to give the proper impulse to the shuttle. The shuttle thus impelled was called the *fly-shuttle*.

The John Kay who invented that fly-shuttle was, according to *The Dictionary of National Biography*, born in 1704 and was "said to have been educated abroad." Since his family seems to have been of limited means, his father having been a clock maker, that education abroad could hardly have been of the kind young gentlemen of family and fortune might then boast, so one wonders if he may not have spent time with relatives whose ancestors had not left the Netherlands in 1585 or France after the revocation of the Edict of Nantes in

1685. Or perhaps they had chosen to move to Germany or Switzerland. Clearly intelligence and manual skills ran in the Kay family.

By 1730, John Kay was established as a reed maker for looms—the reeds being the frames for the vertical sections of cane (or, in our day, metal) which kept in line and separated the warp threads of a loom and served to beat up each weft thread after the shuttle had been thrown. Kay's first improvement was to substitute thin polished blades of metal for the cane. This change made the reeds more durable and better adapted to weave fabrics of a much finer and more even texture.

In 1733, Kay took out a patent for his fly-shuttle, which is often considered the most important improvement ever made in the loom. The woolen manufacturers of Yorkshire were the first to take up this invention with enthusiasm, which did not extend to letting the inventor profit by it. Kay became involved in lawsuits for infringement of patent in which he was successful, insofar as the courts were concerned, but was nearly ruined by the expenses of prosecuting his claims. Added to these problems was the active opposition of workers who saw Kay's fly-shuttle depriving them of work and income. In 1753, a mob broke into Kay's house at Bury, destroying everything they found there, and Kay himself barely escaped with his life. He then sought refuge in France, where he is said to have died in poverty and obscurity while his improvement in weaving machinery was helping make fortunes for manufacturers of all textiles—silks, linens, woolens, cottons. Such was to be the case with most inventors of those times.

A letter from John Kay dated in 1764 runs: "I have a great many more inventions than what I have given in, and the reason I have not put them forward is the bad treatment that I have had from woollen and cotton factories in different parts of England twenty years ago, and then I applied to Parliament, and they would not assist me in my affairs, which obliged me to go abroad to get money to pay my debts and support my family."

5. Thread for Hungry Looms

The real trouble with John Kay's invention was that it demanded thread which could hardly be found for any price. The single-thread spinners had already been having a hard enough time meeting the demands of the old-style cottage looms. Multiplying the speed of looms meant, of course, multiplying the demands, which required either a tremendously increased number of spinners or a greatly increased number of threads that each spinner could produce. The time for spinning machines had come.

The first man to try to fill this need was to be as much neglected and ignored as John Kay. This man was John Wyatt of Birmingham, who devised in 1738 a machine for "spinning by rollers." For a long time, John Wyatt received little credit for his part in this important invention since, because of personal money problems, he did not try to patent it in his own name but used the name of a more solvent associate. Furthermore, Wyatt's invention was presently superseded, if not actually copied, by the man now usually credited as the inventor of the spinning jenny, one Richard Arkwright, who, it is only fair to admit, might have developed the same invention quite independently. It was an invention not only that the times demanded but also that could make a fortune for the man who knew how to exploit it. Richard Arkwright was such a man.

The story goes that the inspiration for the spinning jenny came when a spinning wheel was tipped over by accident and the spindle continued whirling, producing thread in the upright position as it had been in the horizontal. Actually, this was nothing new; Guate-

Note: Unless otherwise indicated, all quotations in this chapter are from Edward Baines, *History of the Cotton Manufacture in Great Britain*, Chapters 8–10.

malan Indians had always spun threads with the spindle held nearly upright. But it seems to have been new to the traditional spinners of Europe and, if the story is correct, suggested to John Wyatt the possibility that a series of upright spindles, all turning at once, could produce a series of threads at one time. This would demand a method for processing raw cotton as it is processed by those Guatemalans but more speedily so that it could be fed into a series of twirling spindles.

The principle back of the spinning jenny and of today's many-threaded spinning machines is basically the same as that described by Baines nearly one and a half centuries ago:

> In every mode of spinning, the ends to be accomplished are, to draw out the loose fibers of the cotton-wool in a regular and continuous line, and after reducing the fleecy roll to the requisite tenuity, to twist it into thread. Previous to the operation of spinning, the cotton must have undergone the process of carding, the effect of which is to comb out, straighten and lay parallel to each other its entangled fibres. The cotton was formerly stripped off the cards in loose rolls, called cardings or slivers; and the only difference between the slivers produced by the old hand-cards and those produced by the present carding engine is, that the former were in lengths of a few inches and the latter are of the length of some hundreds of yards.

This procedure had been managed by having the carded fiber run through a series of paired rollers, or drums, each pair being made to revolve at a speed greater than that of the pair before, so that the sliver can be drawn out to whatever thinness may be needed and thus made ready for the final process in the preparation of thread. After the last pair of rollers, "the reduced sliver is attached to a spindle and fly, the rapid revolutions of which twist it into a thread and wind it upon a bobbin"—all a mechanization of the method employed since time immemorial in remote Guatemalan towns.

"It is obvious," Baines continues, "that by lengthening or multiplying the rollers, and increasing the number of spindles, all of which may be turned by the same power, many threads may be spun at once, and the process may be carried on with much greater quickness and steadiness than hand-spinning. There is also the important advantage, that the thread produced will be of more regular thickness and more evenly twisted."

Neither the historian nor the inventor's patent application gives much space to describing what may have been that "same power" which was supposed to keep the wheels turning. At first it was, quite literally, either mule power or horsepower, the animal of choice being hitched by some unspecified means to a bar sticking out from a drum or wheel which was kept turning by keeping the animal moving in a circle. This drum or wheel was attached by gears and belts to the spinning machine. But neither mules nor horses seemed to move fast enough or to have the necessary endurance to power machines that were to speed up output and replace human labor.

In the eighteenth century, there were few power alternatives. Wind power was erratic, for there was no way to store the power in high winds against the ensuing calms. There remained then only one satisfactory alternative—water power. The water from a flowing stream was backed up in a pond designed to act as a power reserve. So until steam power came of age, it would be water power applied through the agency of huge, lumbering wheels over whose "buckets" water poured as it was led from an upper-level source to a lower level. The same water might be made to perform such a service several times in succession if the downhill course of the stream allowed for several reservoirs, each backing its own wheels. We can see the remains of such water-powered mills where one nineteenth-century factory followed another along the course of a not very imposing stream from which clever and determined individuals managed to extract the last ounce of water power. Today, as in Cohoes, New York, the reservoirs are dry, the channels choked with weeds, yet not sufficiently filled to obscure their original purpose.

John Wyatt, whose business sense failed to match his inventiveness, was not to profit by his spinning machine as he might have had he remained in Birmingham to keep an eye on his machine and its spinning instead of departing for London in search of markets. Years later, in 1817, John Wyatt's son Charles gave an account of the invention and of its fate in a letter he wrote to his younger brother at the latter's request. The account was published in *A Repertory of Arts, Manufactures and Agriculture* "to rescue from oblivion . . . a name dear to us, and unknown to those who are exalted, though perhaps unconsciously so, by his genius: our parent John Wyatt of Birmingham."

Charles Wyatt wrote:

> The brief history of the invention . . . as far as I am able to trace it, is this: In the year 1730, or thereabouts, living then at a village near Litchfield, our respected father first conceived the project, and prepared to carry it into effect; and in the year 1733, by a model of about two feet square, in a small building near Sutton, Coldfield, without a single witness to the performance, was spun the first thread of cotton ever produced without the intervention of the human fingers, he, the inventor, to use his own words, "*being all the time in a pleasing but trembling suspense.*" The cotton wool had been carded in the common way, and was passed between two cylinders from whence the bobbin drew it by means of a twist.

Money ever being in short supply in the Wyatt family, John Wyatt went into partnership with a Lewis Paul, who promised to help with funds but managed only to take out a patent in his own name. Charles Wyatt's account continues:

> In 1741 or 1742, a mill, turned by two asses walking round an axis, was erected in Birmingham, and ten girls were employed in attending the work. Two hanks of cotton then and there spun are now in my possession . . . This establishment, unsupported by sufficient property, languished a short time, and then expired. The machinery was sold in 1743. A work upon a larger scale, on a stream of water, was established at Northampton . . . The work contained 250 spindles and employed fifty pair of hands

. . . The work at Northampton did not prosper. It passed, I believe, into the possession . . . of a gentleman of law in London, about the year 1764, and from a strange coincidence of circumstances, there is the highest probability, that the machinery got into the hands of a person, who, with the assistance of others, knowing how to apply it with skill and judgement, and to supply what might be deficient, raised upon it by a gradual accession of profit an immense establishment and a princely fortune . . .

Such, or nearly such, being the early history of this invention, I thought the late Sir Richard Arkwright would be gratified in possessing the very model to which I have alluded; and I accordingly waited upon him at Cromford with the offer, but my reception did not correspond with my expectations.

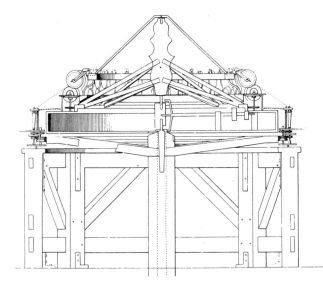

A very mild way of saying that he was snubbed!

What Sir Richard's motives were in rejecting the thoughtful offer can only be conjectured now. He wanted nothing of an invention that some folk, as well as his own conscience, might claim had priority over his own. He wanted nothing of that inventor's son, whose courteous offer he may have suspected hid some devilish scheme to challenge not only his own claims but also the great fortune he owed to his own spinning machine. He might, however, have risked a more cordial response to Charles Wyatt, for the latter, without prompting, was quite willing to allow such claims as Arkwright made, only questioning the complete priority and wishing a bit of credit might be allowed his beloved father's memory.

Long after Sir Richard's death, Charles generously admitted:

> To pretend, however, that the original machinery, without addition or improvement, would alone have produced the prodigious effect which we now behold, would be claiming improbable merit for the inventor, and degrading the talents and sagacity of his successors in the same field of enterprise; for it cannot be denied, that a great fund of ingenuity must have been expended in bringing the spinning works to their present degree of perfection. The number of spindles now in use is supposed to exceed five millions.
>
> If the author of the humble establishment at Birmingham gave birth to such a wonderful progeny, he ought at least to be acknowledged as a benefactor of his country, and recorded amongst the men who, from an attachment to the sciences and the practice of mechanics, open the paths of knowledge, and point out, but do not pursue, those which lead to profit and prosperity.

Richard Arkwright, despite infringement of his own patent rights, was one who had taken those pathways to profit and prosperity, following them from a humble birth and boyhood to a princely fortune and a position of unquestioned preeminence in the textile factory world.

What this all goes to prove is that when the time for an invention has arrived, someone is likely to produce it, but that someone is not necessarily the one to profit by it. With demands for cotton fabrics burgeoning and Kay's fly-shuttle making it possible for looms to spew forth more cloth at a faster rate, the time for a spinning jenny had come.

6. Cotton Creates a Knight

So whom should we honor as the progenitor of the spinning jenny? The long-ignored John Wyatt, perhaps. Or perhaps one James Hargreaves of Blackburn, carpenter and hand-loom weaver residing in an area where carpenters were often called upon to construct looms. Described as "illiterate and a stout, broad-set man about five feet ten inches high," he had a large family whose womenfolk, at least, must have had a hand in his invention. For it is reputed to have been one of their spinning wheels that providentially overturned, suggesting to the man of the house the possibility of having eight spindles turning at once, thereby providing him with all the thread he needed for his own weaving.

Soon, with that large family to feed and clothe, Hargreaves was consenting to build such machines for sale to neighboring weavers—an unfortunate decision since it would help to invalidate later patent claims. The decision was disastrous for other reasons, too, for spinners on single-thread wheels soon realized that their monopoly was being threatened. Mobs formed and headed for Hargreaves' home, gutting it and destroying both his hated machine and his loom. They could not, however, destroy his idea or his skills, which moved with him to another town where he formed a partnership with a man who would build a small cotton mill using the new spinning jenny. This gave the inventor enough cash to pay for a patent—dated July 12, 1770. As inevitably as spinning machines were to become an accepted part of the burgeoning cotton industry, there would arise claims that Hargreaves had appropriated the inventions of others.

Note: Unless otherwise indicated, all quotations in this chapter are from Edward Baines, *History of the Cotton Manufacture in Great Britain*, Chapters 8–10.

The actual fact, however, was that others were infringing upon Hargreaves' rights, making him an entirely legitimate member of the company of inventors.

In 1783, an article published in *Transactions of the Society for the Encouragement of Arts, Manufactures, and Commerce* listed several models of spinning machines which it held in its repositories, for the improvement of which it had been dispensing premiums. Baines stated: "From the best information hitherto obtained, it appears that in the year 1764, a poor man, of the name of Hargreaves, employed in the cotton manufacture near Blackburn, in Lancashire, first made a machine in that county, which spun eleven threads and that in the year 1770, he obtained a patent for the invention. The construction of this kind of machine, called a *Spinning Jenny*, has since been much improved, and is now in so high a degree of perfection, that one woman is thereby enabled with ease to spin a hundred threads of cotton at a time" (p. 154). Like John Wyatt, Hargreaves realized no

fortune from his invention. With his death in 1778, he left behind for his family very limited material resources plus doubts as to whether he was the one who had invented the spinning jenny.

The next in line for such credit was a man who knew how to seize and hold it—Richard Arkwright, whose success and fortune would win him a title. Born in 1732 at Preston in Lancashire, Arkwright was the youngest of thirteen children in an humble if not actually poor family. At an early age, he was apprenticed to a barber. His education consisted of reading lessons given him by an uncle as well as attendance, during winter months, at a local school. His apprenticeship finished in 1760, he set up his own barbering business which, in those days, relied heavily on the wigs he could sell to both men and women. "Shortly after," Baines tells, "he began to travel through the county to buy human hair"—for wigs, of course—"attending for this purpose the hiring fairs frequented by young girls seeking service." No further explanation is given, but one may guess that such young girls, soon to be condemned to wear maids' caps, were quite willing to cut their locks for a price. Beyond this, Arkwright worked out some formula for dyeing the hair, so that he not only produced well-dyed wigs but could add to his profits by selling to other wigmakers hair dyed by his special formula.

Unfortunately, wigs were beginning to go out of style, so the resourceful hair buyer and dyer had to turn his efforts in another direction. It is likely that in the course of his travels from one Lancashire fair to another, Arkwright had heard much gossip about the local textile industry. He must have become aware of that increasingly popular fiber, cotton, and of that new invention that performed the miracle of making many threads instead of only one. Exactly what he was told and how detailed the report may have been, we cannot know. But we do know that by 1767 Arkwright—no spinner by trade—had begun his own work on a "spinning by rollers" machine.

Presently, with the cooperation of one John Kay (no relation to the one of fly-shuttle fame) and a friend, John Smalley, "liquor merchant and painter," the new machine was being set up in the parlor of a house belonging to the Free Grammar School and hidden away behind a garden filled with large gooseberry bushes. The secrecy of the place and the strange goings-on there began to arouse suspicions. There must be witchcraft and sorcery at work here, insisted the old women who had been hearing the strange humming noises that suggested the devil at work tuning his bagpipes for Arkwright and his friends to dance a witches' reel. Somehow Arkwright managed to escape the threatening mob violence to retire, by early 1759, from his Bolton wig shop to devote all his time to perfecting "his" machine for spinning by rollers.

Widowed in 1755, Arkwright in 1761 married Margaret Biggian, who brought with her a dowry of four hundred pounds, which "though settled on herself," must have helped her husband get his start in the new undertaking—a fact which she surely came to regret. Her husband, according to Baines, "was impatient of whatever interfered with his favourite pursuits; and the fact is too strikingly characteristic not to be mentioned, that he separated from his wife not many years after their marriage, because she, convinced that he would starve his family by scheming when he should have been shaving, broke some of his experimental models of machinery." The historian failed to hint whether this breach might ever have been healed. Despite her sacrilegious act, Arkwright and his invention owed much to her four hundred pounds that had subsidized his work. Probably there was no gesture of reconciliation for, as Charles Wyatt was presently to learn, Arkwright was not a man who readily admitted indebtedness of any kind.

Richard Arkwright, in fact, sounds like a character out of fiction—the typical bloated capitalist of a kind few people now really believe to exist. Baines, living within a few decades of Arkwright's life, draws a fascinating and revealing picture of the man:

> I have found myself compelled to form a lower estimate of the inventive talents of Arkwright than most previous writers. In the investigation I have prosecuted, I have been guided solely by a desire to ascertain the truth. It has been shewn that the splendid inventions which even to the present day have been ascribed to Arkwright . . . belong in a great part to other and less fortunate men. In appropriating these inventions as his own, and claiming them as the fruits of his unaided genius, he acted dishonourably, and left a stain upon his character, which the acknowledged brilliance of his talents cannot efface. Had he been content to claim the merit which really belonged to him, his reputation still would have been high, and his wealth would not have been diminished. That he possessed inventive talent of a high order, has been satisfactorily established . . . But the marvellous and "*unbounded invention*" which he claimed for himself, and which has too readily been accorded to him,—the *creative faculty*, which devised all that admirable mechanism so entirely new in its principles and characteristic of the first order of mechanical genius—*this* did *not* belong to Arkwright . . .

The most marked traits in the character of Arkwright were his wonderful ardour, energy, and perseverance. He commonly laboured . . . from five o'clock in the morning till nine at night; and when considerably more than fifty years of age,—feeling the defects of his education . . . he encroached upon his sleep, in order to gain an hour each day to learn English grammar, and another hour to improve his writing and orthography . . . He generally travelled with four horses, and at a very high speed [the eighteenth-century equivalent of a gas guzzler]. His concerns in Derbyshire, Lancashire, and Scotland were so extensive and numerous, as to show at once his astonishing power of transacting business and his all-grasping spirit. In many of these he had partners; but he generally managed in such a way, that, whoever lost, he himself was a gainer. So unbounded was his confidence in the success of his machinery, and in the national wealth to be produced by it, that he would make light of discussions on taxation, and say that *he* would pay the national debt. His speculative schemes were vast and daring; he contemplated buying up all the cotton in the world, in order to make an enormous profit by the monopoly: and from the extravagance of some of these designs, his judicious friends were of opinion, that if he had lived to put them into practice, he might have overset the whole fabric of his prosperity!

However many questions may be raised as to the total credit claimed by Arkwright for the invention of spinning by rollers, the consequences of that invention were certainly largely his, though once such a machine had been invented, such consequences were almost inevitable. The factory system for producing textiles was bound to follow upon the new machines, but it was Arkwright who gave it its first impulse, which eventually led to the establishment of factories in many and varied industries, "with its minute division of labour and regular uninterrupted co-operation of numerous individuals in different processes of machinery. In overcoming the prejudices of workers, in accustoming them to unremitting diligence during the stated hours of labour, in training them for their particular tasks and inducing them to conform to the regular celerity of the machinery, Arkwright displayed an energy and perseverance perhaps of a higher order, if less rare, than that which enabled him to originate his inventions." This execution of so complete and relatively smooth a change in people's outlooks, as well as occupations, was certainly no mean achievement for the man responsible.

Cotton Creates a Knight

The very model of an eighteenth-century capitalist, he looked every inch the part when his services toward improving the economic life of his nation (plus, perhaps, the judicious expenditure of part of his fortune) were rewarded in 1786 by knighthood. Sir Richard Arkwright then sat for his portrait. Judging by that portrait, Carlyle was to describe Arkwright as "a plain, almost gross, bag-cheeked, pot-bellied Lancashire man with an air of painful reflection, yet also of copious digestion." The bulbous-tipped nose in a heavy-jowled face looks out at the beholder across a great expanse of striped vest, which his very wide-open waistcoat could not possibly be made to cover had those pudgy hands strained at it ever so much. The right hand rests on his hip; the other on a table in front of a small cogwheel which, presumably, was to inform the viewer that here sat an inventor of machines, and a highly prosperous one at that—something none of his predecessors in the field of textile machinery had ever been able to claim.

The portrait suggests a supremely self-confident, not-too-scrupulous man who had made his mark in the world, even though he, too, had to fight for what he claimed to be his rights as well as for the credit he believed due him. He had made it easier for his rivals to steal his inventions by overreaching himself in his patent application. For if Arkwright had been ingenious enough to improve upon the inventions of others, he tried to discourage competitors by not revealing all important details in his patent applications. He was to pay for this sharpness when he undertook a suit for infringement of patent against fellow manufacturers who had hurried to equip their own factories with jennies of Arkwright's design without obtaining his permission or paying him a royalty.

In his suit, Arkwright claimed these rivals had "devised means to rob him of his inventions, and to profit by his ingenuity" and that "his servants and workmen (whom he had with great labour taught the business) were seduced" whereby "a knowledge of his machinery and inventions was fully gained." Defendants pointed out that a machine constructed precisely according to the description given in his patent application could not be made to work. That was true, Arkwright admitted, "as the defendants claimed, [but] he had not included in the description of his machine every detail needed to guide others in duplicating it." It was not, as he virtuously insisted, because "he meant a fraud on his country." What he had done had been to confuse men of other countries who might try to build copies, "in prevention of which evil, he had purposely omitted to give so full and particular a description of his inventions, in his specifications, as he otherwise might have done." As his fellow-countryman W. S. Gilbert might have quipped a few decades later, "In spite of all temptations to deal with other nations, he remained an Englishman."

Arkwright lost his suit but not his fortune, which continued to grow to princely proportions, since he owed that fortune less to the jenny itself than to those new establishments of which his jenny became an indispensable part—the great textile factories of Lancashire, Darbyshire, and Scotland.

Of course, Arkwright's jenny was not by any means the last word in spinning machines. Other inventors were breathing down Arkwright's neck, devising their own machines or improving his. With these machines it was possible to spin hundreds of threads, making spinning so effective that threads of the finest quality could be drawn out. By regulating tension and carefully timing the various steps, workers could make the machines almost as responsive as a skilled human hand, yet far more productive.

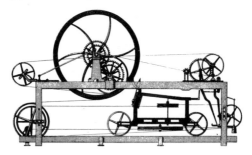

The machine that first made this possible was the "spinning mule," invented by Samuel Crompton, whose sensitive face—the very antithesis of Arkwright's—looks out from a contemporary portrait. Possibly forewarned by the experiences of inventors who had gone before, Crompton made no effort to patent his invention and, in spite of a grant made him by Parliament in 1812, remained a poor man all his life. Crompton's improvement of the jenny made it possible to have machines "of eight hundred spindles each, and some of the prodigious number of eleven hundred." This invention made it possible to produce cotton yarn so swiftly that by 1827, the year Crompton died (Arkwright died in 1792), there were probably at least 50,000 power looms working in the factories of Great Britain.

7. Factories by the Rivers

Whatever questions might be raised as to who was really responsible for the invention of this or that machine, no one questions that it was Richard Arkwright who helped start the factory system—not just textile factories but factories of many kinds—which the new machines demanded. As the numbers of spindles multiplied and spinning machines grew in size, they first crowded the cottages, then the stoutly built workshops needed to accommodate increasingly heavy machines. Finally, even these were not adequate.

William Radcliffe described the transition years in a town near Manchester:

> In the year 1770, the land in our township was occupied by between fifty and sixty farmers, . . . and out of these, there were only six or seven who raised their rents directly from the produce of their farms; all the rest got their rent partly in some branch of trade, such as spinning or weaving woolen, linen or cotton. The cottagers were employed entirely in this manner, except for a few weeks in the harvest. Being one of those cottagers, and intimately acquainted with all the rest, as well as every farmer, I am better able to relate particularly here how the change from the old system of hand labour to the new one of machinery operated in raising the price of land. Cottage rents at that time, with convenient loom-shop, and a small garden attached, were from one and a half to two guineas per annum. The father of a family would earn from eight shillings to half a guinea at his loom: and his sons, if he had one, two or three alongside of him, six or eight shillings each per week; but the great sheet-anchor of all cottages and small farms, was the labour attached to the handwheel; and when it is considered that it required six to eight hands to pre-

pare and spin yarn, of any of the three materials I have mentioned,
sufficient for the consumption of one weaver,—this shows clearly
the inexhaustible source there was for labour for every person
from the age of seven to eighty years (who retained their sight
and could move their hands) to earn their bread . . . without
going to the parish.

From the year 1770 to 1788, a complete change had gradually
been effected in the spinning of yarns; that of wool had disappeared altogether, and that of linen was also nearly gone; cotton,
cotton, cotton was become the almost universal material for employment; the hand-wheels were all thrown into the lumber-
rooms; the yarn was all spun on common jennies; the carding
. . . was done on carding engines . . . but the finer numbers . . .
were still carded by hand, it being the general opinion at that
time that machine-carding would never answer for fine numbers.
In weaving, no great alteration had taken place during these
eighteen years, save the introduction of the fly-shuttle . . .

The families I have been speaking of, whether as cottagers or
small farmers, had supported themselves . . . as their progenitors
from the earliest institutions of society had done before them . . .
But [between 1788 and 1803] an increasing demand for every
fabric the loom could produce put all hands in request, of every
age and description. The fabrics made from wool and linen van-
ished, while the old loom-shops being insufficient, every lumber-
room, even old barns, cart-houses and outbuildings of every description, were repaired, windows broke through the old blank
walls, and all fitted up for loom-shops.

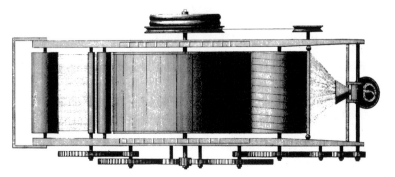

Clearly industrialization was on its way. Baines wrote about the
year 1835:

> Of the fifteen hundred thousand individuals whom the cotton
> manufacture now supports, the greater number are in the county
> of Lancaster. In the year 1700, Lancashire numbered only
> 166,200 inhabitants . . . in 1831 the population had grown to
> 1,386,854; being an increase of more than eight fold in 130
> years . . .
>
> Such are the amazing creations of the cotton machinery. At
> the beginning of the reign of George III (in 1760) probably not
> more than forty thousand persons were supported by the whole
> cotton manufacture: machines have been invented which enable
> one man to produce as much yarn as two hundred and fifty or
> three hundred men could have produced then . . . and the effect
> has been, that now the manufacture supports fifteen hundred
> thousand persons, upward of thirty-seven times as many as at the
> former period . . . It might have been supposed, that the history
> of the cotton manufacture would have forever put an end to the
> complaints against machinery except on the part of the workmen
> who were immediately suffering, as some generally will for a time,
> from the changes in manufacturing processes. The 150,000
> workmen in the spinning mills produce as much yarn as could
> have been produced by 40,000,000 with the one-thread wheel;
> yet there are those who look upon it as a calamity that human
> labour has been rendered so productive. These persons seem to
> cherish secretly the preposterous notion, that, without machinery,
> we should have had as many hands employed in the manufacture,
> as it would require to produce the present quantity of goods by
> the old processes; not considering that the population of all Eu-
> rope would have been quite inadequate to such a purpose.

The various machines, taken together, gave a "mighty impulse" to
cotton manufacture.

> Weavers could now obtain an unlimited quantity of yarn at a
> reasonable price; manufacturers could use warps of cotton, which
> were much cheaper than the linen warps formerly used. Cotton
> fabrics could be sold lower than ever before had been known.
> The shuttle flew with fresh energy, and the weavers earned im-
> moderately high wages. Spinning mills were erected to supply

the requisite quantity of yarn. The fame of Arkwright resounded through the land; and capitalists flocked to him, to buy his patent machines, or permission to use them . . . Mr. Arkwright and his partners also expended, in large buildings in Derbyshire and elsewhere, upwards of £30,000 and Mr. Arkwright also erected a very large and extensive building in Manchester, at the expense of upwards of £4,000. (Baines)

Thus, as Baines noted, a business was formed which Arkwright calculated already employed "upwards of five thousand persons, and a capital on the whole of not less than £200,000." Not bad for the erstwhile maker of wigs!

According to Baines, "The factory system in England takes its rise from this period. Hitherto the cotton manufacture had been carried on almost entirely in the houses of the workmen"—the earlier machines had been placed in home workshops—"but the water-frame, the carding engine, and the other machines which Arkwright brought out in a finished state, required both more space than could be found in a cottage, and more power than could be applied by the human arm. Their weight also rendered it necessary to place them in strongly built mills, and they could not be advantageously turned by any power then known but that of water." Hence came the use of "water frame" to indicate Arkwright's water-powered engine.

The historian continues:

"The use of machinery was accompanied by a greater division of labour than existed in the primitive state of the manufacture; the material went through many more processes; and of course the loss of time and the risk of waste would have been much increased, if its removal from house to house at every stage of the manufacture had been necessary. It became obvious that there were several important advantages in carrying on the numerous operations of an extensive manufacture in the same building. Where water power was required, it was economy to build one mill and put up one water-wheel, rather than several. This arrangement also enabled the master spinner himself to superintend every stage of the manufacture . . . Another circumstance which made it advantageous to have a large number of machines in one manufactory was, that mechanics must be employed on the spot, to construct and repair the machinery, and that their time could not be fully occupied with only a few machines.

All these considerations drove the cotton spinners to that important change in the economy of English manufactures, the introduction of the factory system; and when that system had once been adopted, such were the pecuniary advantages, that mercantile competition would have rendered it impossible, even had it been desirable, to abandon it.

So new and exciting was this change in method and in pace that it all carried an aura of romance. To imaginative individuals, machines that could displace human labor did not seem dehumanizing but rather a glorification of the power of the human mind to control and use matter and thus make it serve human ends. In 1771, after viewing Arkwright's new mill in the River Derwent in action, Dr. Erasmus Darwin, grandfather of the more famous Charles, paid homage to the new trend in a long poem entitled *The Botanic Garden*. In this poem he followed the cotton plant (*Gossypium*) in the progress of its fiber through the factory, using all the ponderous circumlocutions of his age:

Where Derwent guides his dusky floods
Through vaulted mountain and a night of woods,
The nymph *Gossypia* treads the velvet sod,
and warms with rosy smiles the wat'ry god;
His pond'rous oars to slender spindles turns,
And pours o'er massy wheels his foaming urns.

Nor did more down-to-earth people wish to quarrel with machines that could add many desirable things to their lives. "It is impossible to estimate," wrote Baines in the mid 1830s, "the advantage to the bulk of the people, from the wonderful cheapness of cotton goods. The wife of a labouring man may buy at a retail shop a neat and good print as low as fourpence per yard, so that, allowing seven yards for a dress, the whole material shall cost only two shillings and four pence . . . The humblest classes have now the means of as great neatness, and even gaiety of dress, as the middle and upper classes of the last age. A country-wake in the nineteenth century may display as much finery as a drawing-room of the eighteenth."

Not the least romantic aspect of it all, people like Erasmus Darwin felt, was the way the new mills were making use of an old source of power—flowing water. Mill wheels for grinding grain are almost as ancient as grain itself. But the utilization of the same power to set other machines in motion was a relatively new idea, especially as applied in factories. Power was essential, for no human hand or foot was strong enough to keep hundreds of spindles—perhaps even thousands—turning at once. The horsepower first tried, and which today survives as a unit for measuring work, was presently replaced, though the method of transmitting such power long remained unaltered. The central drum turned by the horse, or water power, was connected to wheels which could activate spinning jennies or spinning mules.

Reminiscing later, another inventor described how this activation was achieved: "the spinning mules at that time were generally driven with ropes made of cotton-mill-waste, from a lying [horizontal] shaft in the middle of the room, and over gallows-pullies above the flywheels on each side of the room. That mode of driving was succeeded by belts, which was in every respect much better."

Whatever the machine, whatever the source of power needed to keep it in motion, that power was, for decades thereafter, to be transmitted through a series of gears, pulleys, and belts. Horses, as the

Factories by the Rivers

motive force in factories, became obsolete soon after they were first used, proving to be an expensive kind of power for machines that had to be kept running for hours on end. Even worse, this source of power was unreliable and erratic, for if a horse or a mule decided to stop moving, not even the most skilled mechanic could do much about it.

Arkwright's first mill at Nottingham was powered by horses—and it was his last mill to be so powered. His second mill at Cromford was powered by water. As Baines put it: "During a period of ten or fifteen years after Mr. Arkwright's first mill was built at Cromford (in 1771), all the principal works were erected on the falls of considerable rivers, no other power than water having been found practically useful" (p. 186).

8. Toy into Turbine

By 1790, Richard Arkwright had found another source of power "practically useful," one that came indirectly from water and, owing its effectiveness to the drive of expanding steam, soon helped make obsolete the huge water wheels once so enthusiastically welcomed for the powering of great engines. Incidentally—and sparked especially by the demands of a burgeoning cotton industry—it challenged inventors to find newer, more efficient ways to make available the power being continually wasted as rivers flowed unchallenged to the sea.

Before 1790, invention had been busily at work finding ways to improve those old vertical wheels which were moved by the water of a flowing stream pushing against the wooden vanes, called "buckets," set in the rim of the wheel. The overshot wheel was the first improvement, by which the water impounded behind a milldam was guided through a sluice to fall over the top of the wheel, thus adding the force of falling water to that of flowing water. Mill wheels grew bigger and were sunk in wheelpits. Outgrowing the milldams, they received water not over the top but at a little above middle height, thereby becoming "breast wheels."

Until the year 1844, the breast wheel was considered the most perfect that could be made. People truly believed that progress could go no further.

It was such wheels undoubtedly that long supplied the motive force to most of the huge textile factories that newly mechanized processes were making possible in England and on the European continent as well as in America. In those early days when industrialization was just beginning and the waters of innumerable rivers still ran freely to the sea without being called upon to perform useful tasks for

Toy into Turbine

people, it could not have mattered too seriously that all great water wheels, whatever their design, were wasteful both of water and of energy. Actually, many such wheels were making effective use of less than one-third of the energy that should have been available in the falling water. Even the best wheels wasted nearly half of the possible energy. But no one was measuring such things. No one even tried to measure them, for people did not yet care. Energy was still to be recognized as civilization's greatest and least expendable treasure.

When a millwright was asked to build a wheel more powerful than any as yet known, he could either try designing a more efficient wheel or else he could just build a larger one on lines already established. It was highly unlikely that the usual millwright could think of any way of improving upon a wheel that was "the most perfect that could be made," so bigger and bigger wheels were built with increasingly impressive "horsepowers."

A horsepower—actually nearly half again as much work as it is fair to expect from the average horse—was presently defined as the amount of work exerted in raising a weight of one pound to a height of 550 feet in one second, or, to use more scientific language, 550 foot-pounds per second. Whatever the foot-pounds exerted, there must have been an awesome magnificence in a huge wheel slowly revolving as water splashed over it and as the machinery it powered set to work. No wonder that when, in 1840, a seventy-five horsepower wheel forty feet in diameter was installed at a mine head in Pembrokeshire, England, a large group of onlookers assembled to marvel at what was believed to be the most powerful wheel in the principality.

Yet a scant eleven years later saw the installation in Troy, New York, of a wheel sixty feet in diameter and twenty-two feet thick. This wheel powered machinery that produced spikes and horseshoes—factories by then having become old hat. Sixty years later this machine had become a melancholy wreck. Nevertheless it still remained impressive enough for young engineering students at Troy's Rensselaer Polytechnic Institute, as one reminisced later, to visit it for the inspiration it might give them as well as to supply an opportunity for airing their newly acquired learning before such young ladies as they could persuade to accompany them on Sunday afternoon strolls.

The Troy wheel has now completely disappeared, but there yet survives as a tourist attraction the spectacular wheel of Laxley on the

Isle of Man, just off the English coast. Reputed to have been of 200 horsepower, over 72 feet in diameter and 6 feet wide, it was set in a handsome and imposing masonry structure. Each minute, the pumps powered by that giant wheel raised about 250 gallons of water from a depth of 1,250 feet within the mine. Most people truly believed there could be nothing better.

Still, a few of those engineers of the early and mid 1800s, the mill-wrights, must have asked themselves, as they were called on to repair over and over again the wear and tear on the axles of their lumbering monsters, if there were not a better way. Practical men, and undoubtedly proud of their practicality, they were without any understanding of that relatively new science called hydraulics. They contented themselves with the old empirical trial-and-error method—trying, erring, trying again, but not altering the basic design of water wheels. That achievement was to be forecast by a man whose "Skill in Mechanicks as well as all Parts of Mathematicks and Philosophy" was well known a full century before.

Whether by accident or design, the start of an altogether new kind of water wheel was then begun by the "learned and ingenious Dr. Barker," about whom we can now discover little more, not even his first name, than his curious little scientific toy called a "Barker's mill." From a treatise on experimental philosophy written by a chaplain to His Royal Highness, the Prince of Wales, we get an inkling of the actual invention: "Where there is a Fall of Water not sufficient in Quantity to turn an Overshot Mill, it is possible to make it turn a new invented Mill, the most simple that ever was made . . . Dr. Barker had this thought and communicated it to me. . . . I took the Doctor's Hint and made the following working Model of it, which I shew'd the Royal Society the Experiment of at their last meeting this summer" (Desaguliers).

The working model must have worked and duly impressed the learned members of the Royal Society. Yet Dr. Barker himself soon faded into oblivion, and there seems to be no record as to how "he had this thought" for the little toy that now bears his name. The reaction principle which drove it is the basis of those lawn sprinklers whose movable arms are attached to a stationary central pipe through which water flows under the usual pressure. The forward motion of the water reacts upon the arms to drive them backwards, making them whirl in a direction opposite to that of the water's flow—miniature jet engines.

What special purpose Dr. Barker may have hoped his mill might serve we cannot now know, though we do know that the principle behind it then suggested no serious competition with the popular overshot wheels of his day, which used water power in the only way conservative millwrights could imagine—turning wheels by making water run over and down them.

By the 1850s that situation was beginning to change. Steam engines of James Watt's design began to offer increasingly menacing competition in a power market where water power had reigned supreme. The challenge was all too clear—find a better way to make use of water power or get out of the market. The better way, previously hinted at by Barker's mill, was already at hand—a turbine which, like every new invention, was slow to catch on and had to make its way against a host of skeptics if not downright detractors.

Here, as always, it is hard to understand why those who pride themselves on being practical should be so slow in adopting anything as highly practical as the turbine almost immediately showed itself to be. Compact, relatively inexpensive to build, and usually under six feet in diameter, turbines could maintain efficiency even when submerged to a depth of nine feet, well below the level of ice formations, which so damaged breast wheels. With great or small fall of water, at high or low velocities, and with widely varying quantities of water, even in backwater, those turbines could maintain a remarkable efficiency. There seemed to be absolutely no reason for not going overboard at once for the new invention.

Typically, though, for some time few saw any reason to accept these new "water wheels having a vertical axis called turbines." The first active American convert was Ellwood Morris of Philadelphia, who took his stand in the *Journal of the Franklin Institute*. This challenge was immediately taken up by James Whitelaw of Glasgow, Scotland, who was firmly in favor of an older type of vertical axis machine, one of which happened to be of Whitelaw's own design.

"Mr. Morris," he announced sarcastically in the *Practical Mechanics and Engineers Magazine*, "is, I dare say, a gentleman of talent, and should he take the trouble to study his subject before putting it on paper, may become an honor to his country."

Undaunted, Morris replied in kind, "Mr. Whitelaw makes an assertion which, if it be a fair specimen of the dependence to be placed on his allegations generally, would inscribe him very low, indeed, in the role of truth."

This was an exceedingly formal and long-winded way for two men of science to call one another liars. Yet it had the fortunate effect of stimulating the interest of other men who might have found themselves bored and indifferent had Morris and Whitelaw been in enthusiastic agreement as to the advantages of using turbines as power sources.

Fortunately for the future of such machines, cotton provided a good way for Morris to prove his point. The Rockland Cotton Mills on the Brandywine in Delaware, a textile factory with already highly developed machines, offered the crucial test as, in those days, few other kinds of industrial establishments could. Morris drew up plans for a turbine, persuaded some machinists in Philadelphia to have a try at manufacturing one, then installed it in the cotton mill situated along a stream where the fall was never above seven feet and could drop to half that in times of low water. The new machine worked well, effectively using 70 percent of the total possible power as computed by Morris. This high efficiency, combined with low wear at the pivot (which meant low cost of upkeep), could not fail to impress previously doubting mill owners. In the end, manufacturers had to accept the turbine as the water wheel of the future and of their fortunes. Inevitably, improvements continued to be made, so that by 1870 no fewer than 300 such patents had been issued.

The progressive factories that, in 1844, were willing to try one of the new wheels could not do so by ordering one from a catalogue or even by directing a millwright to reproduce a wheel functioning in a neighboring, possibly rival mill. Wheels were generally contracted for with an eye for the special requirements of the mill in question as well as for the water supply. It was the Appleton Company of the great textile center, Lowell, Massachusetts, that in 1844 ordered for its picking house one of the first turbines to be installed in that state. Designed, installed, and tested by Uriah Boyden, it showed itself all of 78 percent efficient—efficiencies being computed on a scale where 100 percent implied no waste of water, no loss of power in the heat of friction, no slippage of belts.

Boyden seemed convinced that he could improve on the already high efficiency. When two years later he arranged to supply three more turbines to the same mill, he was willing to sign a contract specifying that his pay should depend on the ultimate efficiency of those turbines. If the efficiency still proved to be "seventy-eight percent of the water power expended," he was to receive $1,200 "for his services and patent rights" (an indication that he was continuing to improve the models). "And if the power derived be greater than 78 percent, the Appleton Company to pay me, in addition to the twelve hundred dollars, at the rate of four hundred dollars for every one per cent of power, obtained above the 78 per cent."

Supremely confident, he had inserted no provision as to the kind of payment due should the efficiency fall below the 78 percent. Was it that, as designer of the testing apparatus also, he knew what it was to tell him? The manufacturers must have trusted him, for when an efficiency of 88 percent was reported, they did not hesitate to part with the princely sum of $5,200. Surely the company was well recompensed for its faith in the engineer: eight years later his turbines were still running smoothly and continued to do so until improvements in turbine design made them, too, obsolete enough to be replaced—as were Boyden's methods of water-wheel testing.

As the number of industries requiring power increased, the allotment of a stream's water power raised yet newer problems. If, as with Arkwright's first water-powered mills, they could be placed along the "falls of considerable rivers," there was no problem in building a series of milldams, one above the other and each supplying water to the mill that owned it. But in nonmountainous textile cities like Lowell, Massachusetts, not too far above tidewater, the fall was low and the total quantity of water became of vital importance. The several industries there had to draw upon a reservoir made by building a dam across the Merrimack River at Pawtucket Falls. The water backed up about eighteen miles, creating a lake that covered some eleven hundred acres.

Buying and selling power or the means of producing it must, in those days, have been an infinitely complicated matter. At Lowell, eleven manufacturing companies, with a capital totaling thirteen million dollars, were competing for power, to be measured by the amount of water each factory was to be allowed each day and the height through which that water was to fall. One "mill power"— about 85 horsepower—costing about fourteen dollars per horsepower per year, was estimated in 1877 as only about one-fifth the cost of a horsepower by steam.

To the Merrimack Manufacturing Company, for instance, went 24⅔ mill powers, "each of which consists of the right to draw, for 15 hours per day, 25 cubic feet (155 gallons) of water per second on the entire fall." In all, 139 $^{11}/_{30}$ mill powers were granted to the various factories crowded along the banks of that river. Much of these mill-powers were used on "turbines of a very superior description," the remainder on breast wheels in which some mill owners stubbornly continued to believe.

Lucy Larcom, poet and schoolteacher who as a teenager during the 1830s had worked in a Lowell cotton mill, wrote of this kind of power:

> I never much cared for machinery. The buzzing and hissing and whirring of pulleys and rollers and spindles and flyers around me often grew tiresome. I could not see into their complications or feel interested in them. But in a room below us we were sometimes allowed to peer in through a sort of blind door at the great water-wheel that carried the work of the whole mill. It was so huge that we could only watch a few spokes at a time, and part of the dripping rim, moving with a slow measured strength through the darkness that shut it in. It impressed me with something of the awe which comes to us in thinking of the great Power which keeps the mechanism of the universe in motion.

On the same river, but in New Hampshire and some thirty miles to the north of Lowell, its mills twenty years younger, lies the once-great textile manufacturing city of Manchester, so named by its founders because of a determination to rival, if not outrival, the English Manchester. There, too, water flowing from behind a dam drove a huge wheel in the basement of the Amoskeag Mills. Later a turbine would replace this to power cotton spinning and weaving machines as well as—much, much later—electric generators.

A one-time mechanic for the long-closed Amoskeag, like Lucy Larcom of Lowell, recalled much later the huge wheels in the mills' basements and the fifty-six-inch-wide belt by which power was transmitted to the machines on the floors above. When that belt broke, as it did in the early 1930s, the sheer weight of it presented serious problems to the workmen struggling to mend it. All this was imprinted on the mechanic's memory, though unfortunately he did not mention whether or how electricity was there generated. He did recall, of course, that when the dam itself broke a few years later, the ensuing flood brought to an end the career of the century-old, already enfeebled Amoskeag.

Inside those many cotton factories for which Lawrence, Lowell, and Manchester were long especially famous, power was transmitted during the 1870s from revolving turbine axles, as from the axles of breast wheels, through a maze of gears, shafts, pulleys, belts, and other shafts that might extend from one mill building to another. By this intricate system, all factory machines were powered—pickers for cleaning fibers, spinners of threads, looms where cloth was woven. The elaborate setup was always dizzying to behold, always deafening to hear, occasionally highly dangerous if a worn belt broke and went thrashing wildly about.

How proud those progressive mill owners of the 1870s were of their new water-driven turbines, so much more compact and efficient than the old breast wheels! How shocked those same individuals would have been had they been able to catch a glimpse of the manufacturing world a few decades in the future when their kind of turbines would be as outdated as breast wheels were already becoming. Thanks to still larger, newer, more efficient turbines whirling many miles away, their great-grandchildren would be buying power not in painfully measured mill powers of water, but in metered kilowatt-hours of electricity. Each machine would have its separate motor and, without capitulating to steam engines, mills would be free to locate away from river banks to any areas their owners might think industrially profitable. The 1880s would have come and gone before such a thing came to pass. Perhaps this was all to the good, for the new machines and the new power were to join forces in creating about as many problems as they would help to solve.

9. The Mills Grow Hungry

A factory machine which could do the work of a thousand hands had to have enough raw material on hand to keep a thousand hands busy. As cotton factories grew, demands for cotton fiber soared. For some decades, at least, supply kept pace with demand, but it was a risky situation for mill owners and workers and for the consumers of cotton. So much depended upon a plant that could not grow to maturity in areas where cotton mills were situated.

In the year 1700, nearly one and a half million pounds of raw fiber had been imported into England, most of it coming from the West Indies and from the Levant. By the 1780s, when the impact of the textile factory system just began to be felt, imports in raw cotton were approaching and passing seven million pounds each year, reaching one and a half billion by 1820. Between 1798 and 1861, consumption of textile raw materials in the United Kingdom approximately doubled for wool and flax while becoming twenty-five times as great for cotton. Cotton piece goods exported from Great Britain amounted to about twelve million pounds in 1800, half a billion by 1860.

All the raw cotton to weave those millions of pounds came increasingly from across the Atlantic. The ten bales reaching England in 1784 from the new United States—then so very new that Liverpool customs officials had to be persuaded that the cotton had been shipped from a foreign country and hence might legally be imported in a foreign bottom—had become about one-third of a million bales by 1820, over two and a half million by 1860. Importations from all other countries put together ranked far behind, the United States then supplying easily two-thirds of all the cotton consumed by those voracious mills of Lancashire.

It was, of course, the Southern states that produced that fiber, their agriculture expanding to meet the mills' burgeoning demands. Georgia's fiber yield of six pounds per capita in 1790 had increased tenfold by 1800, fiftyfold by 1860. South Carolina's production followed much the same pattern. These amounts could not have been produced save for the inventiveness of a Northerner—neither a slaveholder nor a cotton planter—who was a visitor on a Southern plantation during the 1790s.

The problem had lain in the particular species of cotton. Sea Island cotton, long-stapled and much in demand, grew only near the coast. Its fine soft fibers could be separated from the smooth black seeds by a rather crude machine that fed the fiber between rollers spaced so close together that the hard black seeds were held back. But the remainder and greatest proportion of the Southern cotton crop was Upland cotton, whose sticky green seeds could not thus be separated. They had to be picked out by hand. A slave woman could produce no more than a pound of cleaned cotton each day—which, even though the worker was an unpaid slave, made Upland cotton an unprofitable crop.

To maintain the standard of living that Southern landowners had come to expect, a really profitable crop was needed. In the aftermath of the American Revolution and the desolation wrought in Southern states, the last to see active fighting in that war, plantations were understaffed and without funds. Impoverished planters dreamed of fortunes to be made if only someone could show them a way to clean Upland cotton fiber. Georgia's and South Carolina's six pounds per capita of 1790 could have increased very slowly had it not been for the machine invented by the visiting Northerner, Eli Whitney.

It was mere chance that had placed the resourceful New Englander in Georgia at the crucial time. Born in Westborough, Massachusetts, just ten years before that "shot heard 'round the world" reverberated from a Lexington hardly more than thirty miles from the elder Whitney's farm, Eli grew up doing farm chores when he had to, but always preferring odd jobs that challenged his ingenuity and inventiveness. Working, studying as he could, hoarding his earnings, Eli managed to enter Yale's freshman class in 1789. There he continued to earn what he could by repairing apparatus for science professors and tutoring such undergraduates as might be in need of help. By such means, he worked his way through to graduation in the autumn of 1792.

It was then that chance took a hand in his—and incidentally cotton's—affairs by making him acquainted with a visiting Yale graduate, a man of about his own age, whose moderate means had enabled him to enter college at a younger age than Eli. This was Phineas Miller, tutor to the children of the late Nathanael Greene, the patriotic Rhode Island Quaker whose vigorous defense of the South had won him the affection of Georgians and South Carolinians to the point where they rewarded him with estates confiscated from departed Tories. The general had died in 1786, and his widow was making her home in Georgia, trying, with the help of tutor Miller, to keep the estate going profitably.

Shortly after his Yale commencement, Whitney was introduced to Phineas Miller by Dr. Ezra Stiles, president of Yale. Eli was then looking for suitable employment and Miller knew a Carolina family who was looking for a tutor for the children. Correspondence followed, what appeared to be satisfactory arrangements made, and young Whitney headed south, accompanying the Greene party when it sailed from New York for Savannah.

About a year later, Whitney wrote to his father describing the events that followed. That letter, dated September 13, 1793, and sent from New Haven, Connecticut, ran in part:

I went from N. York with the family of the late Major General Greene to Georgia. I went immediately with the family to their

Plantation about twelve miles from Savannah with an expectation of spending four or five days and then proceed to Carolina to
take the school as I have mentioned in former letters. During
this time I heard much of the extreme difficulty of ginning cotton, that is, separating it from its seeds. There were a number of
very respectable gentlemen at Mrs. Green's who all agreed that if
a machine could be invented that would clean cotton with expe-
dition, it would be a great thing both to the Country and to the
inventor. I involuntarily happened to be thinking on the subject
and struck out a plan of a Machine in my mind, which I communicated to Miller (who is agent to the Executors of Genl. Greene
and resides in the family, a man of respectability and property)
he was pleased with the Plan and said if I would pursue it and
try an experiment to see if it would answer, he would be at the
whole expense, I should lose nothing but my time, and if I succeeded we would share the profits. Previous to this I found I was
like to be disappointed in my school, that is, instead of a hundred, I found I could get only fifty Guineas a year. I however
held the refusal of the school until I had tried some experiments.
In about ten days I made a little model, for which I was offered,
if I would give up all right and title to it a Hundred Guineas. I
concluded to relinquish my school and turn my attention to perfecting the Machine. I made one before I came away which required the labor of one man to turn it and with which one man
will clean ten times as much cotton as he can in any other way
before known and also cleanse it much better than in the usual
mode. This machine may be turned by water or with a horse,
with the greatest ease, and one man and a horse will do more
than fifty men with the old machines. It makes the labor fifty
times less, without throwing that class of People out of business.
(cited by Bruchey, pp. 60–61)

It would not have occurred to the inventive genius that what his
"gin" (short for "engine") was to achieve was the fixing of a pattern
of cotton plantation slavery which previously had been thought to be
languishing.

Whitney's letter continues:

I returned to the Northward for the purpose of having a machine
made on a large scale and obtaining a Patent for the invention. I
went to Philadelphia soon after I arrived, made myself acquainted

with the steps necessary to obtain a Patent, took several of the steps and the Secretary of State, Mr. Jefferson agreed to send the Patent to me as soon as it could be made out—so that I apprehend no difficulty in obtaining the Patent. . . . As soon as I have got a Patent in America, I shall go with the machine which I am now making, to Georgia, where I shall stay a few weeks to see it at work. From thence I expect to go to England [to obtain a patent there] where I shall probably continue two or three years. How advantageous this business will eventually prove to me, I cannot say. It is generally said by those who know anything about it, that I shall make a Fortune by it. I have no expectation that I shall make an independent fortune by it, but I think I had better pursue it than any other business into which I can enter. Something which cannot be foreseen may frustrate my expectations and defeat my Plan, but I am now so sure of success that ten thousand dollars, if I saw the money counted out to me, would not tempt me to relinquish the object. I wish you, sir, not to show this letter nor communicate anything of its contents to any body except my Brothers and Sisters, *enjoining* it on them to keep the whole a *profound secret.* (cited by Bruchey, p. 62)

Almost a year later a still hopeful Eli was writing his father from New Haven:

It gives me pleasure that I have it in my power to inform you that I am in perfect health. I left Savannah just three weeks ago. We had a passage of Eight Days to New York, where I spent several days and have been here about a week. I was taken sick with the Georgian fever about the middle of June and confined to my bed ten or twelve days, but had got quite well before I left the Country. There were several very hot days preceeding my sickness, during which I fatigued myself considerable and which was probably the cause of my illness.

My Machinery was in opperation before I came from Georgia. It answers the purpose well, and is likely to succeed beyond our expectations. My greatest apprehensions at present are, that we shall not be able to get the machines made as fast as we shall want them. We have now Eight Hundred Thousand weight of Cotton on hand and the next crop will begin to come in very soon. It will require Machines enough to clean 5 or 6 thousand wt. of clean cotton pr Day to satisfy the demand for next Year. I mean for the crop which comes this fall. And I expect the crop will be double another year. (cited by Bruchey, p. 62)

It had been a futile gesture, in the earlier letter, to enjoin secrecy upon his father and sister and brothers. They lived in a state where no cotton grew and people couldn't care less how fiber destined for its textile mills had been cleaned in the state where it grew. The unkeepable secret was on its way out as soon as Whitney proudly put his first model to the test. In Georgia, once the existence of such a machine became known, planters came from long distances to view the invention which just might turn Upland cotton into a profitable crop and thus enrich all of them.

An early historian of the cotton gin described the situation:

Their impatience could not be restrained. The shed which contained the cotton-gin was forcibly entered and in the morning the machine was gone. The principle of its construction—as yet unpatented—was discovered. New machines, with slight and unimportant variations, were manufactured and set up in various parts of the state. The owners of the original gin (Mr. Whitney had taken as a partner Mr. Phineas Miller, who had married his friend, Mrs. Green) were involved, after the issue of their patent in the fall of 1793, in almost endless litigation. Their rights, moral and legal, were shamefully disregarded.

In spite of the loss of their only model, and the infringement of their patents, Whitney and Miller still had hopes of securing a share of the wealth which their machine was sure to create. Their plan was to sell no machines but to gin cotton for the planters on shares, the owners of the gin retaining one pound in every three. This turned out to be an unfortunate plan. Whitney, who went North for the purpose was unable to supply the needed machines. The scarcity of money, due to wild speculation in land, crippled his operations. Scarlet fever broke out among his workmen; and to cap a long series of misfortunes, just as Whitney was recovering from a serious illness, he arrived in New Haven to find his factory and his half-finished machines in ashes. This was a serious blow. Not only was the financial loss large, but the impatient planters, who had raised an immense quantity of cotton, the value of which depended upon its being ginned, were given extra inducements to make machines for themselves in spite of the patent.

The owners of the cotton gin were not disheartened by their misfortunes. They raised money at ruinous rates of interest, and proceeded with their enterprise. (cited by Bruchey, pp. 56–57)

Meanwhile, spiteful rumors were being spread about Whitney's gin—that it damaged the fiber and made it less saleable or, at least, reduced the amount that would be paid for the ginned fiber. Until time proved this false, the partners saw their gin standing idle in a cotton-growing country. Nevertheless, planters—possibly even those so busily engaged in running down Whitney's machine—were actively involved in the infringement of the partners' patent, their rights being totally disregarded. Individual infringers could be located, but it proved to be quite another thing to make a successful case against such, most of whom were wealthy and influential and socially above a mere inventor. Nevertheless, Whitney stuck by his guns and brought the first case for patent infringement in 1797. On prejudice, rather than on technicalities, it was decided against the firm of Whitney and Miller. No less than sixty such cases were to be presented before a verdict could be secured in favor of the claimants. This was in 1808, when the patent had only one more year to run and after the resources of the concern had been exhausted. Whitney's philosophical summing up might have been applied with equal justice by those English inventors of spinning and weaving machines whose rights had also been shamelessly disregarded. As Whitney saw it, his troubles were due to "the want of a disposition in mankind to do justice," adding, "I have always believed, that I should have had no difficulty in causing my right to be respected, if it had been less valuable, and been used by a small portion of the community" (Bruchey). Against the many prominent infringers, he was helpless. He had, nevertheless, one advantage over his equally unfortunate British counterparts—no one has since questioned that to Eli Whitney must go the credit for a machine which "unlocked the imprisoned resources of the South" (cited by Bruchey), for whatever such financially unrewarding fame might be worth.

Whitney can hardly be blamed for the far-reaching consequences of his invention. It was to change the lives and occupations of great numbers of his countrymen, transform the previously sluggish life of the South into one of activity, power, and wealth. Incidentally, it gave new stimulus to the institution of black slavery, thus feeding the political developments which were to culminate in the Civil War. And across the ocean, it was to involve the lives and fortunes of many, from the factory owners in Lancashire to the workers in their factories and to the peoples of the various lands from which that war's embargo was to force those factory owners to purchase the raw cotton they needed.

10. Water into Steam

No such far reaching effects could have materialized in a process relying on manpower or horsepower or even on water power, for water power as applied in the late 1700s had one obvious drawback. It had to be applied close by some flowing stream. All those startlingly new machines that were filling people of those times with wonder—the spinning jenny, the fly-shuttle, and the cotton gin—had in common that basic requirement for an outside source of power. They all had had to follow the same path from hand crank through horsepower to water power and beyond, as factories grew in size and complexity and the water of flowing streams, no matter how hoarded, was reaching a limit. Once land sites along stream banks were all occupied and all water allotments were contracted for—what then? The possibility of ever using the newly recognized "electric fluid" to power anything was then hardly dreamed of, so invention had to come to the rescue in the late eighteenth century with the first practical steam engine. Steam then seemed as endlessly available as the fuel—wood or coal—to produce it seemed inexhaustible.

The story of the steam engine and its inventors has an all too familiar ring. Like those other inventions more intimately connected with the production of textiles, it is full of frustrations and stolen ideas, with final fame going to those who, if worthy, were hardly more worthy than those who died in poverty. It was an idea whose time had come, even though it did not come to all who deserved it.

For women, long aware of the steam issuing from the spout of a teakettle or raising the heavy lid of an iron pot, it may have seemed strange that inventors were so long in recognizing the power of steam. Yet in days when servants were plentiful, few inventive men would have ventured into their kitchens where pots might be boiling

60

or remained there long enough to observe what was happening and draw lessons from these mundane phenomena.

The first English engine which made use of the expansive power of steam owed its development to the economic necessity of its inventor, the Marquis of Worcester. Born in 1601, the Marquis was a wealthy and idealistic nobleman who, during the English Civil War, gambled both freedom and fortune on the wrong man—the greedy and fickle King Charles I. Having spent lavishly in a vain effort to keep Charles upon his throne and Charles' head upon his shoulders, the Marquis was bound to be looked upon with suspicion by Oliver Cromwell, Lord Protector of the commonwealth which had engineered Charles' downfall. So the Marquis had to spend the years from 1650–1656 as a prisoner in the Tower, where, if nothing else, he had plenty of time to think and plan.

In 1660, when the second Charles ascended the throne, the Marquis cheered himself with the hope that the son might remember and perhaps repay the man to whom his father, quite literally, owed so much. In 1661, the Marquis submitted a "statement of the Marquis of Worchester's expenses for King and country." By then, time had given him a juster estimate of the gratitude of kings, and he ended his statement cynically: "These sums added together balance the accounts and make good that I have spent, lent (and lost?) for my king and country £918,000" (Dircks, p. 334).

Even by today's rate of exchange, that sum would amount to over one and a half million dollars. In those days it was the equivalent of considerably more.

With negligible repayment forthcoming even after the Restoration, the Marquis had to live by his wits—and he was fortunate to have more than average wits to live by, though unfortunate in having them lead him far ahead of his times. All his life he had amused himself by inventing things, some of them curiosities of no particular practical value, some of them useful little gadgets. He managed to keep in his employ a skilled mechanic whose duty it was to assemble and keep in repair the machines his employer invented. The Marquis was especially proud of one such machine—a "water-commanding" engine to "raise water by fire" to a tank on the roof of his castle, whence it might flow down to whatever spot the Marquis desired and keep a fountain in his courtyard constantly playing.

That device was hardly what we of today would call an engine. It had no moving parts and worked by using the pressure of steam in a boiler to force water up into a halfway tank. From here it could be forced still higher by suction produced in an adjoining tank when the steam already in it was allowed to condense. Thus, with two halfway tanks, each acting alternately as a pressure and a suction chamber, joined through appropriate valves, the marquis could raise water to as great a height as anyone in his day could dream of desiring.

When, at long last, the Marquis had to face the fact that noble monarchs could be very ignoble when it came to settling debts, he concentrated all his inventive efforts upon developing and demonstrating his "stupendous water-commanding engine" through which he expected to recover part, at least, of his squandered fortune. Somehow he managed to purchase the house and grounds of Vauxhall, across the River Thames from London, and there set up his engine where possible purchasers might view it, marvel, and place orders. Visitors did come in numbers, though only to view the fountains powered by the Marquis' amazing engine, which was talked of far and wide. Two years after the Marquis' death, the engine must still have been working, for the Florentine duke Cosimo de Medici made a special point of seeing Vauxhall while visiting London in 1669. Perhaps the duke, recalling his own father's problems of over thirty years past when he could find no suction pump capable of pumping water from a newly dug, very deep well, was better able than most Londoners to appreciate what the Marquis of Worcester had accomplished. There is, however, no record that he coveted such an engine for his own use.

The Marquis had died a poor and disillusioned man, his engine ignored save as a curiosity. Otherwise it was practically forgotten until, in 1699, with the Marquis' patent long since run out, it was revived as the invention of Thomas Savery, Captain of Engineers. Since we can now locate no accurately detailed records of the Marquis' engine, it is very hard to reach a just decision as to whether the captain deliberately stole the Marquis' ideas, as implied in a book of 1744, or whether it was really a sort of accidental copying by a man who had heard much talk of a wonderful engine once performing at Vauxhall. Today it cannot matter deeply where Savery picked up his ideas. What does matter is that, by 1698 and 1699, Savery had managed to acquire two patents for steam engines couched in such general terms as to tie up the whole steam engine business for the next thirty-five years. Certainly this was Savery's greatest stroke of genius.

Savery's engine, like the Marquis', had no moving parts except the valves controlling access of steam and water. Also like the Marquis' engine, it was of limited use. But at about the time Savery acquired his patents, Thomas Newcomen, an ironmonger—that is, hardware merchant—was developing his own engine that was to remain the model for all pumping engines for nearly seventy-five years. Though all his engine had in common with Savery's was the use of steam for motive power, it was covered by those general terms of Savery's patent. Newcomen had to enter into a partnership with Savery and to share his profits. From start to finish, Savery, as Richard Arkwright was to show himself to be a century later, had a genius for striking it rich in other men's inventions.

The Reverend Desaguliers, who despised Savery, wrote somewhat less carpingly about Savery's enforced partner: "About the year 1710 Thomas Newcomen, ironmonger, and John Calley, glazier . . . made then several experiments in private, and having brought [their engine] to work with piston &c in the latter end of the year 1711 . . . they bargained to draw water for Mr. Black of Wolverhampton, where, after a great many laborious attempts they did make the engine work, but not being either philosophers to understand the reasons, or mathematicians enough to calculate the powers and to proportion the parts, very luckily, by accident they found what they sought for." It did not seem to occur to the writer that the so-called "accident" could not have happened without the seeking, even though by men less "philosophers" than the revered members of the Royal Society.

Improvement in engines was eventually brought about by a man of whom Desaguliers would have approved, since he had those qualities of philosopher—or "scientist"—and mathematician the peppery chaplain thought so important. This was James Watt. Born in 1736 in Scotland, apprenticed at eighteen to a London instrument maker, the young man soon grew homesick and moved back to Glasgow, where in 1757 he became mathematical instrument maker for the university. There he was bound to become well acquainted with the men who were using those instruments, among them Joseph Black, professor of chemistry at the medical college of the university. Incidentally, medical colleges were then the only institutions where basic sciences like chemistry and biology were studied.

Black happened to be especially interested in heat and its profound effect upon chemical reactions. More specifically, he was fascinated with what we now call latent heat of vaporization—the heat, for instance, which must be put into boiling water to keep it boiling though no thermometer registers any rise in temperature. Thus the term "latent" for that heat which, as most of us have learned to our sorrow, does not remain latent if our skins are exposed to live steam, the resulting scalds being much more severe than those caused by boiling water.

Black completed his famous research upon heat before he moved on, in 1766, to a professorship at Edinburgh. Meanwhile, James Watt built the instruments the professor ordered, surely learning along the way much about heat and steam. In later life, as a highly successful builder of Watt steam engines, he acknowledged his abiding personal indebtedness to Black and to this work on heat.

By 1763, Watt was attempting to become financially independent enough to get married—which was not likely on the pittance paid a mere instrument maker. He set himself up as a general engineer, in which work he soon acquired a considerable reputation. Clients engaged him for a wide range of undertakings: making surveys and estimates, supervising public works such as canals, harbors, and bridges. These were before the days of engineering schools when engineers learned by doing and clients judged and employed them on the basis of what they had done.

Never too busy to do favors for his old university associates, Watt agreed, during the academic year 1763–1764, to repair and put into running condition an old demonstration model of Newcomen's atmospheric engine. For Watt, this proved to be the kind of challenge he enjoyed dealing with. If a model didn't work, why didn't it? On the other hand, why had it ever worked? Since it was the suction caused by the condensation of steam that pulled on the piston (as steam inside a sealed can pulls upon the sides of the can when it condenses), conditions had to be regulated so that the steam behind the piston condensed at precisely the right time and in the right place.

Watt soon perceived that in the Newcomen model it was not just the steam but the whole cylinder that cooled during the condensing phase of the cycle. Thus the piston had to be reheated and the water revaporized each time, with an accompanying wastage of fuel that even in those days of plentiful fuel could not be ignored. Moreover, more energy could be made to use not only the condensing suction but also the expanding pressure of steam—as, of course, had the unfortunate Marquis of Worcester's water-commanding engine.

Thus stimulated, Watt's mind went to work and by 1769 he was ready to patent his own steam engine, which incorporated improvements based upon careful research—a second and separate cylinder for condensing the steam, a steam jacket for keeping the main chamber hot, and steam so used that it pushed as well as pulled upon the piston. He further invented devices commonplace in our times but startlingly new then, such as a centrifugal governor to keep engine speed within safe limits in addition to a float to control the water level inside the boiler.

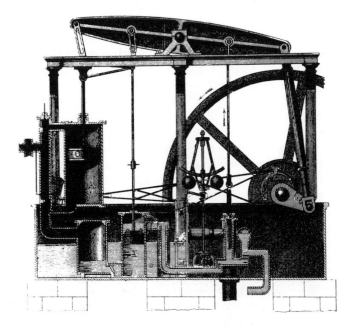

So thorough and painstaking was Watt's work that the engine was soon doing its own promotion. Contemporary amazed comment makes it sound like one of our twentieth-century mechanical brains:

In the present perfect state of the engine, it appears a thing almost endowed with intelligence. It regulates . . . the number of its strokes . . . counting or recording them, moreover, to tell how much work it has done, as a clock records the beats of its pendulum. It regulates the quantity of steam admitted to work . . . the supply of water to the boiler; it oils its joints; . . . and when anything goes wrong which it cannot itself rectify, it warns its attendant by ringing a bell . . . it is the king of machines, and a permanent realization of the *Genii* of eastern fable, whose supernatural powers were occasionally at the command of man.

Genii, perhaps, but of the English teapot rather than of the Oriental lamp.

Neither at the time of the invention of that wonderful machine nor for a considerable time thereafter could most people see any reason for trying to improve upon its supernatural powers. In any case, Watt together with his partner and financial backer Matthew Boulton—of the firm Boulton & Watt—had taken care of all that with a patent that was to keep the steam engine business entirely in their own hands until 1800. The firm prospered. The partners grew wealthy although all that they demanded from mine owners who replaced old engines with their new ones was the equivalent of one-third of the saving in fuel affected by the change.

Here, mine owners must have told themselves, is the end of the road—the topmost peak of efficiency from which we can aspire no higher. But a new and younger generation of engineers was rising who could not believe any machine so perfect that it was beyond improvement. One of these was young Richard Trevithick, a generation younger than Watt, who believed that the use of steam under high pressure—eight atmospheres or 120 pounds per square inch—would improve considerably the efficiency of such engines. However, until the expiration of Watt's patent in 1800, there was not much Trevithick could do about it, for, with or without exclusive patent rights, Watt was still the accepted prophet of steam power and no young Cornish upstart was going to displace him.

By then, the brilliant, open-minded, warmly friendly engineer Watt had become the successful businessman with a stake in keeping things just as they were. Watt may have perhaps convinced himself that he was acting the part of a concerned humanitarian when he tried to get Parliament to forbid, in the cause of public safety, the use of "murderous" high-pressure steam engines, as a contemporary described them.

Competition with a firmly entrenched business proved too much for the young Cornishman, who lacked any personal financial backing. Yet he was in many ways even more a prophet of the future than James Watt. The year 1801 saw him constructing a steam-propelled locomotive which was actually able to draw a load on Christmas Eve of that year. Even this startlingly new contraption roused no spark of response in his conservative countrymen, opened for him no new opportunites. After having to try his luck in Peru and Costa Rica during the years from 1814 to 1827, he returned home to petition Parliament in vain for some financial aid. That petition denied, he died penniless in 1832.

Trevithick's tragedy was one shared by every inventor who sees too far beyond the times. He saw steam as a wider source of power for both locomotives and ships. He saw ships as abandoning their old types of wooden hulls for iron ones. What he lacked was people to share his vision. Only in 1870, many years after Trevithick's death and when iron ship hulls had become a commonplace, was a high-pressure steam engine patented by the American George Corliss.

Meanwhile, an increasingly power-hungry age was demanding more power than old-fashioned piston-and-cylinder reciprocating mechanisms could produce. Inventors, who by now had learned how to compute the efficiency of such engines, turned to a new idea which was, in fact, an old idea in a new-fashioned form, powering machines with a turbine driven by steam. Actually, it was a very ancient idea indeed, for as early as the second century before Christ, so the legend goes, Hero of Alexandria invented a little hollow ball provided with two curved arms out of which steam could be made to issue. The force of the issuing steam drove the arms backwards, as does the water in today's lawn sprinklers. This Alexandrian rotating ball is said to have been used to wind up on an attached drum the ropes which controlled the opening and shutting of temple doors. When an acolyte lit the altar fire in the morning, the heat would cause the water to boil. The steam thus produced made the ball turn and the temple doors open. At night, when the altar fire flickered out, the doors would close, making the superstitious exclaim, "Behold how our gods keep watch over their own!"

Whoever may have been responsible for the idea—Egyptian gods, or Hero of Alexandria, or perhaps some other long-forgotten Egyptian prophet-inventor—the invention had to wait twenty centuries to be adapted for running powerful machines directly or, later, indirectly through electric power transmitted across long distances. Either way, steam was to take a hand in powering cotton factories, intruding upon the lives of thousands—owners and workers and just average citizens of the lands to which they belonged or with which they did business.

The powering of cotton factories by steam engines had certainly been encouraged in Lancashire by the fact that the fuel lay underfoot. Historian Sir Arthur Arnold wrote in 1864: "All the larger cotton towns, with the single exception of Preston, are situated upon, or in close proximity to the strata known as coal measures." Had the coal not been there, Lancashire, "instead of producing clothing for millions upon millions of human beings, might have echoed to the crow of the grouse or the tinkle from the bell-wether of the mountain flock." A pretty picture, but Sir Arthur seemingly forgot that for many decades Lancashire had been clothing millions with the aid of its rivers and their breast wheels, if not yet turbines.

11. King Cotton's Expanding Realm

For one thing, at least, King Cotton had to thank the quantities of coal so easily available in Lancashire. Due to the very young chemical industry, the coal gas it was producing by 1806 could practically double the production of factories while minimizing the dangers previously inherent in lighting by candles—by compressing the light of [twenty-three] candles into a single jet at less than a quarter the cost. Thus the hours of factory labor, the owners gleefully perceived, could be stretched from well before dawn to well after sunset.

While horses and asses that had once turned machines were put out to pasture, the workers at those machines were becoming increasingly enslaved. Transferring the work of a hundred hands to a single machine may, furthermore, have speeded up production of cotton textiles, as had the cotton gin through the more rapid cleaning of fiber, but in the end these advances only complicated the problems planters, factory owners, and workers had to face. Aware that their continued prosperity depended upon acres planted in faraway lands like India or Egypt or the United States, mill owners of Lancashire lived on the verge of panic lest crops fail and their factories be forced to grind to a halt. Such are the risks of becoming dependent upon the ruler of any plant kingdom.

Each alternative source of cotton had its drawback. The United States being no longer a British colony, its planters had to be begged rather than ordered to produce more and better cotton. Yet India, though technically under British dominion, was more distant in spirit as well as in location. There, until the Mutiny of 1857 brought its power to a total end, the English East India Company long dominated cotton production.

The chief problems were basic with the plant itself. Had all cotton

69

fiber come from a single species of the plant genus *Gossypium* and had all growers produced fibers of equal desirability in the constantly increasing quantities demanded by Manchester factory owners, there would have arisen few problems for the factories beyond the obvious one of transporting raw material over thousands of miles of ocean. Yet none of the most serious problems were immediately evident.

In the decades after the disruptions of the Napoleonic Wars came to an end, cotton supplies continued to increase to meet the ever-rising demands of the trade. It was a trend that Lancashire manufacturers came to rely upon—that their needs would continue to be filled. Only vaguely aware of what it could mean to be totally dependent upon the vagaries of a plant's growth, they ignored the possible disastrous effects of floods, droughts, pests, or infecting organisms. They barely concerned themselves at all that the superior American crop, which was coming increasingly to displace cotton from other lands, depended upon a population of slave workers that was also on

the increase. The social, political, and agricultural dangers inherent in the situation did not, as viewed from across the wide Atlantic, seem to hold much threat for mill owners concerned only with keeping themselves supplied with the raw material they needed. They had always had the cotton they needed, and, despite expressions of concern by a few farsighted individuals, they expected to keep muddling along indefinitely.

Perhaps manufacturers of cotton goods in England took it for granted that cotton fiber from a British-controlled India could always fill any gap that might arise should a source outside the empire fail them. After all, cotton growing and spinning and weaving had flourished in India since before recorded history. Early travellers in the Far East had been moved to comment on the fineness and beauty of the cotton fabrics they encountered in India.

Two Arabian travellers of the ninth century recorded: "In this country they make garments of such extraordinary perfection, that no where else are the like to be seen. These garments are . . . wove to that degree of fineness that they may be drawn through a ring of moderate size" (quoted by Baines, p. 56). One of the early Portuguese adventurers who visited the Far East, always with an eye to profitable commerce, noted "the great quantities of cotton cloths admirably painted" (quoted by Baines, p. 56). A French traveller named Tavernier commented more interestingly, during the middle seventeenth century, on the "calicuts"—that is, calicoes or muslins—of India. They were "woven in several places in Bengal and Mogulistan, and are carried to Raioxsary and Baroche to be whitened, because of the large meadows and plenty of lemons that grow thereabouts, for they are never so white as they should be till they are dipped in lemon-water. Some calicuts are made so fine, *you can hardly feel them in your hand*, and the thread, when spun, is *scarce discernible*." This was all to the good since, "when a man puts it on, *his skin shall appear as plainly through it, as if he was quite naked*; but the merchants are not permitted to transport it, for the governor is obliged to send it all to the Great Mogul's seraglio and the principal lords of the court, to make the sultanesses and noblemen's wives shifts and garments for the hot weather; and the king and the lords take great pleasure to behold them in these shifts, and see them dance with nothing else upon them" (quoted by Baines, pp. 57–58). Clearly, clothes had come a long way since those concealing coverings of fig leaves in the Garden of Eden!

An English writer of the same period, obviously worried about future competition of such fine muslins with the sturdy products of his homeland, complimented the Indian muslins, though not intentionally, by titling a 1696 pamphlet dealing with such matters *The Naked Truth*. "Fashion," he wrote, "is truly termed a witch; the dearer and scarcer any commodity, the more the mode; 30s. a yard for muslins, and only the shadow of a commodity when procured" (quoted by Baines, p. 78).

Prices must have soon dropped, for about twelve years later author Daniel Defoe, in his *Weekly Review*, was lamenting the national evil of preferring cheap clothing from abroad to homemade commodities at a higher price:

> The general fansie of the people runs upon East India goods to that degree, that the *chints* and *printed calicoes*, which before were only made use of for carpets, quilts, &c, and to clothe children and ordinary people, become now the dress of our ladies; and such is the power of a mode as we saw our persons of quality dressed in Indian carpets, which but a few years before their chambermaids would have thought too ordinary for them: the chints was advanced from lying upon their floors to their backs, from the foot-cloth to the petticoat: and even the queen herself at this time was pleased to appear in China and Japan, I mean China silks and calico. (Defoe, vol. 4, p. 606)

"This time" was not in 1708, when Defoe's article was published, but a decade earlier, before "the prohibition of Indian goods had averted the ruin of our [woolen] manufacturers, and revived their prosperity," in Defoe's words. The Act of Prohibition, passed under William III (of Orange), forbade the introduction of Indian silks and printed calicoes for domestic use, either as apparel or furniture, under penalty of a £200 fine for the wearer or seller. It could not, of course, prevent the smuggling of such goods, though anyone found wearing them might have a lot of explaining to do.

Explanations, however, would not long be necessary, for the act was underwriting the infant textile industry of Lancashire. By 1831, with textiles from British factories now flooding the Bengal markets, the natives of that faraway land were rather pathetically petitioning the Lords of His Majesty's Privy Council for Trade:

Your petitioners most humbly implore your Lordship's consideration . . . and they feel confident that no disposition exists in
England to shut the door against the industry of any part of the
inhabitants of this great empire.

They therefore pray to be admitted to the privilege of British
subjects, and humbly entreat your Lordships to allow the cotton
and silk fabrics of Bengal to be used in Great Britain "free of
duty" or at the same rate which may be charged on British fabrics consumed in Bengal. (quoted by Baines, p. 82)

The confidence thus expressed was misplaced, for, though the petition sounds altogether reasonable from the perspective of a century
and a half, it was denied. Manufactured Manchester cottons and
their manufacturers were in the saddle, or at least they thought they
were. Yet their remaining so depended upon a plant that could not in
any way be persuaded to mature in a climate like England's, and mature it must if those factories were to keep on producing.

It seems hardly fair that the same group who fought to retain the
duties on Indian cotton fabrics should have expected to continue re-
ceiving bales of raw cotton fiber from the same land. But at the time
they fought for those duties, they were seeing India only as a rival
and were not really relying on a crop which was making up only a
small part of the total amount being consumed in the mills of England and Scotland. Ironically, they were relying on Indian markets
as consumers of the goods their mills were producing.

Three decades later, the shoe was on the other foot as the American Civil War slowed to an almost invisible trickle the export of cotton from the states of the South, once the suppliers of about four-
fifths of all cotton fiber consumed in the British Isles. And even that
trickle was contraband that might at any time dry up altogether.

12. Tributary Lands

When the blockading of the ports of the Confederate states by Union forces threatened to shut off altogether the shipping of cotton from the South, the same factory owners who had so labored to bring about the interdiction of imported Indian cotton textiles expected that the gap between their supply and their needs for raw cotton would be filled by cotton fiber growing in British-controlled India. It was then that they began to learn the drawbacks of such a source.

Whatever kind of fiber might be grown there, a product which was sold and bought in native bazaars by weight invited adulteration by native middlemen who, perhaps with some justification, couldn't care less about the factories destined to use it. When bales of Indian cotton reaching those English factories were opened, sticks, dirt, and even stones were often found. They not only added to the apparent weight but also complicated the problems of putting that cotton through temperamental machines. Moreover, Indian cotton was short-fibered and contrasted sadly with the highly desirable long-stapled Sea Island variety, which everyone agreed was the best the states of the American South produced. Even American Upland cotton was better than most of the Indian cotton that reached England. Worst of all, Indian cotton could not be ginned by the kind of machine that served so well on American plantations.

When the years 1846 and 1847 turned out to be short crop years in the American South, Lancashire factory owners began to get a foretaste of the worse disasters that were already threatening. The need to find several alternative sources of raw cotton fiber was being dramatically brought home to the capitalists who knew little of tropical agriculture. It was in this area that their sad ignorance of the botanical facts of life was to serve them in very bad stead. They began to ask the seemingly obvious question: "Why not plant American seeds in Indian soil so that a failure of the crop in America might just be compensated for by a fine crop of the same cotton on the opposite side of the globe?" They found it nothing short of stupid that the best American cotton had not been planted—or, at best, was not growing—all over India. Such "practical" people must have laughed with scorn at the notion that the shape of the root systems of different species of cotton decides the crop's success or failure.

Most native Indian cotton varieties, having developed in areas where rainfall was confined to a single monsoon season each year, survived because they were endowed with tap roots shaped like long, thin carrots, which could reach down into the soil far enough to draw upon deeper sources of water. American Sea Island cotton had a shallow, spreading root system, fine for the very moist Sea Island climate but offering no insurance against a very dry season. In India, it could grow only in the limited area which enjoyed two monsoon seasons a year. Such was Dharwar in the Bombay presidency, where Sea Island seed, once planted, was to produce the much-valued variety called "Dharwar-American."

One thing the Manchester factory owners could not ignore was that no matter how fine the quality of the fiber raised in India, it was of little profit until it reached their mills. And it could not reach their mills from the interior of India, where they fancied huge stocks to be piling up, unless there was some way of getting it to the coast and then loading it onto ships for the long voyage to England.

Roads were bad. Railroads were all but nonexistent in India, there being no more than 332 miles of track in all of India in 1858, when the government of India was finally transferred from the East India Company to the Crown. Port facilities were pitifully inadequate. Yet all such facilities, plus improved river transportation, had to be made available if bullock carts were to deliver baled cotton to railheads or riverboat landings where those bales could be sent to ports capable of handling ships and of loading cargoes expeditiously. Decades were to pass while the not too well informed factory owners of Lancashire struggled to persuade the scarcely better informed Parliament and the Government of India to bring about the much-needed improvement of transportation facilities. In that faraway land where the better means of communication was bound also to affect the lives of native millions, the politics of cotton was becoming very complicated indeed.

Transportation problems were not the only ones responsible for the piling up of cotton stocks in India. There were still older problems, thanks to the short-sighted policy of that curious survival of Elizabethan days when East India Companies were being formed, first in England, then in continental lands, to systematize and protect commerce with the newly accessible Far East. All-powerful in seventeenth-century India, but not satisfyingly solvent as Company factors in India and investors in Britain saw it, the English East India Company missed no chance to increase its funds by taxing local commodities. As demand for cotton fiber soared in England and the cultivation of that crop was pressed in India, taxes were to be collected before the bales left India. Taxes required inspectors who must inspect and assay the crop once it had been harvested. Once a crop had been harvested, it could not legally be moved until an inspector had looked it over. With such officials taking their time, cotton bales awaiting inspection were stored not in sanitary sheds but in dung-lined, straw-covered pits, where deterioration as well as that inevitable contamination with sticks, stones, and other foreign materials that native middlemen felt no impulse to remove combined to lessen the value of the product. Buyers of Indian cotton were finding it even dirtier than American slave-produced cotton which, by comparison, was thought of as clean and pure.

Other possible sources of raw cotton included Egypt, where cotton was an ancient though never much exploited crop. There the problems were different. If, as the Manchester factory owners seemed to feel, cotton growing in India suffered from neglect by those in authority, cotton growing in Egypt suffered from the opposite tendency, though only recently so. India had, for many centuries, been producing, for export as well as for home consumption, its own textiles from cotton grown in many separate areas. In contrast, Egypt's short-staple cotton remained a minor crop until, in the nineteenth century, a reigning pasha became suddenly and acutely aware of that fibre's potential as a source of income. He brought his mighty powers of persuasion to bear upon everyone from peasants bound to the land to middlemen, largely foreign, who dealt in the commodity.

The pasha's gesture was not necessarily a totally self-seeking one though admittedly his indirect efforts to improve the incomes of those peasants, called *fellahin*, may have been an attempt to give a totally impoverished and despairing peasantry some incentive to keep on working and, of course, paying taxes.

In the last analysis it was really the Nile river that determined what crops were grown in Egypt and when, as it had been doing since before recorded history. Herodotus of Halicarnassus, writing in the fifth century before Christ, was deeply impressed with the importance of that river to the whole Egyptian economy. He described the river's changing moods, as might any historian of any age, before or since:

> The Nile divides Egypt in two from the Cataracts to the sea, running as far as the city of Cercasorus [Cairo] in a single stream, but at that point separating into three branches . . . Meanwhile the straight course of the stream, which comes down from the upper country and meets the apex of the Delta, continues on, dividing the Delta down the middle, and empties itself into the sea by a mouth . . . Now the Nile, when it overflows, floods not only the Delta, but also the tracts of country on both sides of the stream . . . in some places reaching to the extent of two days' journey from its banks.
>
> Concerning the nature of the river, I was not able to gain any information either from the priests or from others. I was particularly anxious to learn from them why the Nile, at the commencement of the summer solstice, begins to rise, and continues to increase for a hundred days—and why, as soon as that number is past, it forthwith retires and contracts its stream, continuing low during the whole of the winter until the summer solstice comes round again . . .

Concerning this problem, Herodotus' fellow countrymen had offered three explanations, two of which he was, with reason, moved to reject out of hand. Of the third, he admitted, it "is very much more plausible but is positively furthest from the truth . . . it is that the inundation of the Nile is caused by the melting of snows . . . how is it possible that it can be formed of melting snow, running as it does, from the hottest regions of the world into cooler countries?" And how could one answer that to a man unaware that anywhere in the world there could exist mountains so high that, even in those hot equatorial lands, snows might remain unmelted on their summits all the year round? Wisely, Herodotus gave up worrying about that question.

"Let us leave these things, however," he wrote, "to their natural course, to continue as they are and have been from the beginning.

With regard to the *sources* of the Nile, I have found no one among all those with whom I have conversed, whether Egyptians, Libyans, or Greeks, who professed to have any knowledge." He would have had to live another twenty-three centuries or so to learn the answer to that question, which was to become an obsession of exploration-minded nineteenth-century Englishmen.

The thing about the Nile most important to our subject—the changing water level which in our century has been controlled through the great Aswan Dam—was recognized by Herodotus, as it had been recognized fourteen centuries earlier by an Egyptian monarch Herodotus referred to as King Sesostris:

> The king then . . . proceeded to make use of the multitudes whom he had brought with him from the conquered countries, partly to dig the huge masses of stone which were moved in the course of his reign to the temple of Vulcan [apparently a fortress] partly to dig numerous canals with which the whole of Egypt is intersected. By these forced labors, the entire face of the country was changed; for whereas Egypt had formerly been a region suited both for horses and carriages, henceforth it became entirely unfit for either . . . being cut up by canals which are extremely numerous and run in all directions. The king's object was to supply Nile water to the inhabitants of towns situated in the mid-country, and not lying upon the river.

Herodotus was not an agriculturalist and did not seem to grasp the agricultural implications of such a canal system.

Those periodic fluctuations of the level of the Nile were responsible for two types of crops, each determined by the season and the water then available to it. Before 1800, about one-eighth of the cultivated area of lower Egypt, which included the lands of the Nile Delta, was used for the so-called "summer" crops, which actually grew during the months of the European winter, to be harvested after the summer solstice when the Nile had begun to rise. Among these crops was a quite limited amount of short-staple Egyptian cotton, which was the only cotton cultivated in Egypt until 1819, when chance revealed a finer long-staple native type. A "summer" crop avoided the river's annual floods as the "winter" crops—sugar, rice, indigo—did not. These required the building of protective embankments similar to the levees along the Mississippi River. During the months of the European winter, the shrunken Nile flowed between such embankments, its level ten to twelve feet below the tops.

It was in such a period that canals such as those built by King Sesostris' chain gangs came into active use. They were supplied with river water through devices as old as Egyptian agriculture itself. One of these—the *shaduf*—was a water wheel turned by oxen, if the landholder was prosperous enough to own draft animals. This wheel, reversing the process by which water wheels of other lands produce power, brought river water up to a level where the buckets that studded the rim could pour the water into a sluice or canal. Expensive in ox power, the *shaduf* also required round-the-clock human drivers if water-demanding crops like indigo or cotton were to be served. The other device was totally manual—the *sadiya*, a modified version of the well sweep that once graced many a colonial dooryard. Water was lifted from the river in a bucket attached to the end of a long pole which then, with the help of a counterpoising stone weight, was swung up to empty the water as did the water wheels.

This combination of man power and ox power persisted well past the middle of the nineteenth century, for there was no other native source of energy. Steam engines were imported to a limited extent during the 1830s, largely for spinning and weaving factories. However, without a native resource such as coal to produce the steam to keep them running, they became very expensive to operate and were not used where the old forms of labor were to be had on demand. Furthermore, with no trained native-born engineers, the engines which had worked intermittently during the 1830s were soon breaking down altogether, making the Egyptian agriculture of the nineteenth century A.D. look not too different from what it had been in the nineteenth century B.C. Cotton would have had a part in that agriculture, though how large a part it is hard to conjecture. Certainly it persisted from before the time of Herodotus. In the year 1586 A.D. Laurence Aldersey, visiting Alexandria and Cairo, described the former as a cotton mart: "The towne standeth in a valley and along the water side pleasantly. There are about 26 windemils about it, and the commodities of it are cotton wool, cotton yarne, mastike and some other drugs."

The cotton growing in lower Egypt—the Nile Delta which extends between Alexandria and Cairo—was generally an annual shrub producing an inferior fiber, as was the fiber from the perennial bush of upper Egypt. Workers on cotton plantations were *fellahin*, grouped under a headman and subject to rigid rules as to who might grow what where, who might set up a factory, what products were to be produced, where they might be sold, and at what price. Limited literacy contributed to the practical bondage of the *fellahin*.

Europeans conducting business in Egypt had to contend with this system while they themselves remained under disabilities that had improved little since the days when Alexandria was a center of a trade in spices brought thither from the Far East for eventual sale in Europe. A 1780 visitor reported that the French residents in Cairo were "shut up in a confined space, living among themselves without external communications; they even dreaded it and went out as little as possible, to avoid the insults of the common people, . . . and the violence of the Mamluks who forced them to dismount from their asses in the middle of the streets."

Mohammed Ali was appointed viceroy of Egypt under the Ottoman Empire in 1805. In 1811, he became total hereditary ruler, in fact, if not in name, and remained so until 1848, when, because of his mental derangement, his son Ibrahim became regent. Though ruthless and illiterate, the father had been shrewd enough to perceive the value of foreign progress in mechanical invention and industrial method and what these might mean to Egypt. It was he who began removing some of those disabilities endured by foreigners whose cooperation he desired, especially in raising and selling cotton crops and in establishing textile factories. Yet only in the 1850s, under Mohammed Ali's grandson, Abbas, were foreigners (and, of course, their investments) encouraged to the point where they dared erect Christian churches.

One foreigner imported by Mohammed Ali was the young French textile engineer named Louis Alexis Jumel, engaged in 1817 to come to Egypt to supervise one of the new spinning and weaving mills. Sometime within the following two years, Jumel spotted, in a Cairo garden, a cotton bush whose fine fibers seemed superior in both length and strength to any then generally known in Egypt. Actually, it was not a totally new plant sport, for a few Egyptian women had already been harvesting and spinning such fiber. But it was new to the textile world at large.

Aware of what such a fiber might do for the industry he had been engaged to assist, Jumel planted in his own garden seed from the newly discovered bush. The year 1820 saw a yield of three bales of Jumel's long-staple cotton. A year later, with financial help from the pasha's government, the yield was two thousand bales of a cotton whose value in Europe was from two and a half to four times that of the more usual Egyptian short-staple cotton. By 1822, Mohammed Ali was excited enough by the Frenchman's find to order that the new cotton be cultivated on an extended scale, allotting to this crop the most easily irrigated lands. He also had *sakiyas* erected, animals sold to peasants on credit, seed provided, and cotton gins as well as presses for packing fiber into bales manufactured and distributed to the villagers. To make assurance of success doubly sure, knowledgeable instructors were engaged.

It was that plant with the desirable soft fluff that was bringing about in a predominantly Moslem land some tolerance of foreigners, even Christian foreigners, whose life there had previously been barely tolerable. Jumel's cotton was providing a direct stimulus to the introduction of spinning and weaving factories where, after the initial importation of a few jennies and looms, machines copied from those importations by Egyptian carpenters and smiths were being installed. During the two years that followed 1824, twelve cotton factories were built, with four bleaching establishments to process the products of the looms. Progress, however, had to stop with power. Gins and jennies and looms were still relying upon human and animal muscles for motive power, to the point where a street peddler in Cairo was heard announcing his cloth with the cry, "The fruit of the bull, O maidens!"

The machine age did not take hold there too successfully at once, as, perhaps, it might have if Egypt had then had fuel resources that could liberate the textile industry from old agricultural and technological methods. The quality of the finished cloth produced by the new factories was low. The machines soon broke down, for neither workers nor managers dared stop them for repairs, since to do so was to risk punishment if they failed to meet the output targets set by superiors who lacked technological understanding. So the machines were kept working until, for all practical purposes, they fell apart.

Mohammed Ali himself understood too little of technical details, but he had become sufficiently aware of the general lack of such knowledge in Egypt to begin, in 1826, sending young Egyptians abroad for education. During the next two years, 108 students received education abroad—69 in strictly technical subjects. Perhaps a worse handicap for the ruler was that there were too few officials upon whom he could rely to carry out the reforms which he felt were needed but which he himself perhaps too little understood. His son

Ibrahim was equally responsive to Egypt's needs. In 1845, Ibrahim's own princely estates boasted thirty-two locally manufactured cotton gins (of a kind especially adapted to the cotton of Egypt) turned by eight oxen, and he was planning to have them replaced shortly by twenty-four American roller gins. These gins were probably to be powered by steam engines since the new, barely recognized source of power, electricity, was not to produce functioning motors of any kind for several decades. Nor, until our own century, with the discovery of oil in the East and the exploitation of turbine-generated electricity made possible by the great Aswan Dam at the Nile's first cataract, could Egyptians run machines without importing fuel as well as engines.

In an 1862 report to the American Secretary of State, the American consul stationed in Egypt, William S. Thayer, stated:

> The principal article of export is cotton [valued at seven million dollars]. The exports to America from Egypt are chiefly rags and gums. The imports are chiefly machinery, furniture and ice. Nearly all the machines used in the cleaning of cotton have been imported from the United States, as well as a considerable portion of the rolling-stock used upon the Viceroy's railway to Suez.
>
> The natural impediment to direct commerce between the United States and Egypt is obviously the identity of the principal exportable productions of the two countries—cotton and grain.

There would, of course, have been little competition in the matter of ice, which must have been a costly and valued commodity. And in the year 1862, when this report was made, any cotton that reached Europe from America would have been contraband.

This was a situation which the Crimean War of 1854–1856 had taught Egypt how to exploit. During this period the peasants received "fabulous prices for all commodites," prices having tripled in two years. Proprietors, wary of increased demands from the Egyptian government, either buried their capital or used it in ways that made the income unavailable to tax collectors. "No fellah," Thayer's report continues, "ventures to appear to possess money. It would expose him to being robbed and beaten by a hierarchy of oppressors" from the headman immediately over him to the pasha himself.

With the outbreak of the American Civil War and the embargo imposed by the government in Washington on all shipments from the Southern states to any destination, notably to England, Egyptian cotton tried to rush in to fill the void. Within the ensuing five years, the weight of Egypt's total annual exports of cotton was multiplied by five, the cash value of those exports by a factor of nearly ten. And this in spite of seriously adverse conditions in Egypt's cotton growing areas.

The first crisis was due to a cattle plague referred to as "murrain" and apparently not, in those very early days of bacteriology and pathology, further identified. Those sciences were, however, already sufficiently advanced for educated owners of cattle to have become aware of how infections spread and to attempt isolation of districts where the murrain was rampant, to forbid moving cattle from one village to another, to cancel village fairs where peasants and their cattle might congregate, and to insist that, instead of following the age-old custom of throwing dead diseased cattle into the river, their carcasses be burned. Meanwhile oxen, horses, and donkeys purchased abroad began arriving, but these were to serve only briefly in powering the machines of the cotton industry. They had been too weak to start with, acclimatized badly, and were immediately overworked. In all, it was estimated that about 70,000 animals perished as a result of that murrain.

The second crisis came with floods caused by the highest Nile waters of the century. Yet in spite of these handicaps, the cotton crop and the prices it brought continued to increase. Egypt was by way of becoming one of the world's large suppliers of cotton fiber. By thus taking an important place at the courts of King Cotton, Egypt was, without having planned it that way, being forced into long-overdue social, educational, and technological advances which, in turn, were to have worldwide repercussions.

13. American Tribute

Thirty years before the start of the American Civil War shook the throne of King Cotton to its foundations, Egypt and the West Indies had each been supplying about one-ninetieth of the cotton fiber consumed annually by the mills of Britain. From the East Indies—largely India—and Brazil had come shipments amounting to about one-ninth of the total consumption each, leaving to the United States roughly four-fifths of the British market. Yet cottons had been growing from time immemorial in the other tropical and subtropical Americas. Asian cottons described by early travellers and/or traders were white or off-white, though one Chinese cotton—nankeen—was known for its light yellow hue. American cottons showed more tints. Of these, a lovely golden brown is still to be encountered in the *huipiles* of certain Guatemalan towns while other colors—notably blues and greens—have been reported in the past.

Unfortunately, few of the earliest travelers in the Americas thought it important to mention the colors of the fibers they saw growing. The conquistadores had more martial matters on their minds. But those enterprising sixteenth-century British merchants who were presumably Catholic enough to be allowed access to Philip II's overseas empire did comment briefly on what they saw. One of these merchant-travelers—John Chilton—left an account in Hakluyt's *Principall Navigations*. Of the Indian's clothing, he wrote: "They are clothed with mantels of linnen cloth, made of cotton wooll, painted throughout with works of divers fine colours." This did not, of course, mean painting in the usual sense but the working in of colored threads. Of Guatemala, he said, "All the Indians of this province pay their tribute in mantels of cotton wool, and cochinelio [cochineal, a prized bright red dye] whereof there groweth abun-

84

dance throughout this countrey." But of the actual colors of the cotton fibers themselves, he contributes no more information than did the Spanish writers.

As an article of trade, cotton was not then stirring travelling English merchants to enthusiasm. Gold and jewels were as much their object as the Spaniards'. An interesting sidelight on John Chilton's travels comes in the barely missed confrontation with his less Catholic fellow countryman, Francis Drake.

"Neere to this place," Chilton wrote, "there lieth a port in the South Sea, called Aquitula, in which there dwelleth not above 3 or 4 Spaniards with certaine Negroes, which the King mainteneth there: in which place, Sir Francis Drake arived in the yeare 1579 in the moneth of Aprill, where I lost with his being there about 1000 duckets which he took away, with such other goods of other merchants of Mexico, from one Francisco Gomez Rangifa, factor there for all the Spanish merchants that then traded in the South Sea."

The account of this episode given by the historian of Drake's voyage runs somewhat differently:

> The next harbor therefore which we chanced with, on April 15 in 15. deg. 40. min. was Guatulco so named of the Spaniards who inhabited it, with whom we had some entercourse, to the supply of many things which we desire, and chiefly bread &c. And now having reasonably, as we thought, provided our selves, we departed from the coast of America for the present: but not forgetting, before we gate a-shipboard, to take with us also a certaine pot (of about a bushell in bignesse) full of ryalls of plate, which we found in the town: together with a chaine of gold, and some other jewells, which we intreated a gentleman Spaniard to leave behind him, as he was flying out of towne. (Fletcher)

Aquitula is clearly today's Acapulco, by the latitude given. It was the last of the west-coast Spanish towns Drake was to visit, for better or worse, on that voyage. The Spaniards were hot after him, and he wisely sailed north for the coast of what, two centuries later, was on the brink of becoming one of the United States.

The distance to be traversed, as well as the bulkiness of such merchandise, long made it unlikely that cotton to feed the hungry mills of Lancashire should be imported from the Pacific coast of America. But within four years of Drake's departure from Acapulco, other Englishmen were already looking toward the east coast of Brazil for that commodity. In that year, an English merchant, appropriately named Edward Cotton, was recording instructions to a captain of a ship he was sending thither: "At your coming to the Isle of S. Sebastian, upon the coast of Brazill, you shall according to your discretions, make sale of such commodities as you may thinke will be there about well vented, and likewise buy commodities without making longer stay than your victuals bee providing, but rather to bespeake commodities against your returne from the River of Plate, especially of Amber, Sugar, Greene Ginger, Cotton Wool, and some quantities of peppers of the countrey there." Nothing ever came of all this, the ship foundering off the African coast with all but four hands lost.

São Sebastião is an island off the Brazilian coast a little to the north of the port of Santos, in whose neighborhood cotton can mature. One would like to know how much cotton was growing there in the late sixteenth century, whether there were cotton plantations of any considerable size, and what eventual destination was projected for the cotton wool Edward Cotton's ship was to have loaded as cargo. Baines rejects any suggestion that cotton yarns were thus early being spun and woven in Manchester, ascribing the occasional mention of spun "cottons" to a corruption of the word "coatings," which must have been fashioned of wool and hardly imported from a largely tropical country like Brazil.

There is no question, however, that cotton could grow and thrive in Brazil, not only in the vicinity of Santos but far to the north, nearer the equator in the states situated on the Brazilian bulge. The problem of these states has always been the reverse of Egypt's—too little water which, to make matters worse, arrived in unpredictable quantities at unpredictable times. The searing droughts of northeastern Brazil and the human misery they bring have long been legendary. Only great dams for impounding water of rivers swollen by the occasional heavy downpours, as projected by government agencies, could hope to supply the water needed to make cotton a crop there which the mills of other lands might rely upon. Though in 1791 Brazil was credited with producing twenty-two million pounds of raw cotton, that figure had increased only by a little more than one-half to thirty-six million pounds by the year 1860, when the world's largest supplier was on the brink of lapsing from competition.

In the course of those same seventy years, the annual cotton production of the United States had soared from a mere two million pounds to sixteen hundred and fifty million pounds, or from about one two-hundredth to about two-thirds of the total world production, most of which was being consumed in the mills of Lancashire, England, though the manufacturers of New England had already begun to breathe down the necks of those of Old England. The clouds of the social and political storm raised by this situation in cotton-producing and cotton-consuming lands had for some time been gathering on the horizon, in general unrecognized. The factory system, slavery, expanding commercial horizons in other lands—all these were adding up gradually to a radical and irreversible change in which King Cotton was to become inextricably involved.

By the 1960s, the picture of world cotton production had changed materially from that of a century earlier when cotton from the southern United States outdistanced all rivals. Still the world's largest producer of raw cotton—as reported by the foreign agricultural service of the United States Department of Agriculture—the United States was, in 1960, producing only one-third of the total cotton of the world, while the production of the southern Soviet Union had risen to one-sixth, with China and India following at about one-tenth each. The United Arab Republic, Mexico, and Brazil were then

each accounting for about one-twentieth of the total world production, Pakistan one-thirtieth, Turkey one-fortieth. Still lower on the production scale were Iran, Nicaragua, Peru, Colombia, Greece, Uganda.

The consumer picture for cotton yarns and textiles had changed still more radically. The United Kingdom had lost its preeminence, with the United States now taking first place in quantity consumed, followed by the Soviet Union, China, and India. Back in 1790, about twenty-five million pounds of raw cotton were being processed in the mills of the world, which were than almost exclusively British, while at the same time single-thread wheels were consuming perhaps twenty times more. By 1910, or shortly thereafter, mill consumption had risen to ten and a half billion pounds annually. Hand spinning was so negligible a quantity as not to figure in the statistics. During the 1880s, those mills in the United Kingdom were responsible for about 82 percent of the international trade in cotton yarns and textiles. This had dropped to 52 percent between 1910 and 1913. By 1966–1968, the United Kingdom's share of the world trade in cotton piece goods was a mere 3 percent. Exports from the United States had declined almost as sharply while Asian countries were picking up the slack. Ironically India, whose native cottons had started it all nearly two centuries before, had become again a major exporter of cotton fabrics. A further irony was that England, having earlier all but destroyed Indian textile manufacture, was again a major importer of Indian cotton goods.

A country that became the world's largest grower and exporter of cotton fiber was bound to become involved in textile manufacturing sooner or later. Baines had noted in 1835: "The growth of cotton manufacture in America has been rapid. The first cotton mill was erected in Rhode Island by 1791" by an English mechanic who, having worked in an English textile mill before emigrating, constructed from memory a mill for spinning cotton.

"As late as 1807," Baines wrote, "there was not in the Union more than 15 mills, producing about 300,000 lbs. of yarn in a year. The embargo of 1808, the differences with England, and above all, the war [of 1812] gave a great stimulus to the manufacturing interest, and led the Americans to indulge the desire of supplying themselves with the cottons and woollens their population desired. High protecting duties were therefore established, which forced the growth of manufactures. In 1810, the number of cotton mills had increased to

102, and in 1831 to 795." Between 1800 and 1815, "the quantity of cotton worked in the year by the United States" had increased from 500 to 90,000 bales.

In 1831, a report of the Committee on Manufacturers of the American Congress stated that in the 795 mills, there were about one and a quarter million spindles, 33,500 looms, all consuming nearly seventy-eight million pounds of cotton fiber each year and producing nearly sixty-eight million pounds of spun yarn. There were 18,539 male workers in these mills and 38,927 females.

Clearly the infant industry was growing fast, but its guardians felt that it needed the protection of those "high protecting duties" mentioned by Baines. It was over such duties that the producer South and consumer North were, already during the 1840s, becoming bitterly divided. Another report of the same congressional committee, released in 1842, included a petition signed by over fifty prominent citizens and manufacturers of New England—"citizens of Boston and the vicinity, interested in cotton manufacture, in contemplation of the proposed revision of the tariff" (Ware). It stated: "The cotton manufacture as an important branch of American industry, takes date from the year 1816, under the specific or minimum duty of that year, and the introduction of the power loom . . . Its rapid extension had been without parallel in the whole history of commerce . . . To estimate its importance in quantity, it is only necessary to observe, that the present consumption [of fiber] is equal to the whole export of the United States up to the year 1820, or the whole consumption of American cotton in Great Britain up to the same period, and exceeds our export to France previous to the year 1840" (Ware). With the South having an almost single-crop economy which thrived on overseas consumption of that crop and the North trying to compete with a long-established English product, the congressional debate on whether to protect home-woven fabrics by a tariff was already foreshadowing the grimmer events to come two decades later.

Cotton fabrics had long since outgrown their early luxury status. Their production, which had expanded beyond the wildest dreams of the inventors and manufacturers of the century before, called for expanded markets. Of the status of the English cotton manufacture in 1833, Baines wrote that there were nine and one third million spindles and one hundred thousand power looms, powered either by steam (equivalent to 33,000 horsepower) or water (11,000 horsepower).

The journalist in historian Baines surfaced in his proud words: "It may help to form a conception of the immense extent of the British cotton manufacture, when it is stated, that the yarn spun in this country in a year would, in a single thread, pass round the globe's circumference 203,775 times; it would reach 51 times from the earth to the sun. . . . the wrought fabrics of cotton exported in one year would form a girdle for the globe, passing *eleven* times round the equator." Income from that manufacture was supplying about two-thirds of Britain's public revenue. Even without putting its globe-girdling potential to the test, that bit of fluff from around the seeds of a semitropical plant had come a long, long way!

14. The King's Servants

Cotton yarn may have helped increase national income and made good fabrics available at a reasonable price to the citizenry at large, but its most important influence was on the people who were involved in producing the new fabrics as they experienced the move from a cottage industry to factory employment. All too uncomfortably aware of this, Sir Edward Baines wrote:

> We have seen the effects of the cotton manufacture, in increasing the commerce, population, and wealth of the kingdom, and in adding to the personal and domestic comforts of all classes. The philanthropist and the moral philosopher will, however, inquire what is the physical and moral condition of the vast population employed in the manufacture? The workmen who constructed or attend upon all these machines are not to be confounded with the machines, or their wear and tear regarded as a mere arithmetical question. They are men—reasonable, accountable men; they are citizens and subjects; they constitute no mean part of the support and strength of the state; on their intelligence and virtue, on their vices and degradation, depend in a considerable measure not only the character of the present age, but of posterity; their interests are as valuable in the eyes of the moralist, as those of the classes who occupy higher stations . . .
>
> The principal considerations will be, the *command which the working classes have over the necessaries and comforts of life,* their *health*, their *intelligence*, and their *morals*.

Factory employment having grown from practically nothing during Sir Edward's lifetime, he was bound to view it all with divided loyalties—loyalty to the new class of factory owners whose existence had come to mean so much to so many Britons as well as loyalty to those working classes who kept the factories going and of whose needs the violent political upheavals of the previous century had made other classes uneasily aware. Factories, in spite of the romanticizing of men like Erasmus Darwin, could be dehumanizing, disruptive of family life in a way we of today can hardly appreciate.

To be just, we should consider how it all got started. The factories, appallingly efficient for their day, were displacing a rather casual cottage industry where several family members worked at spinning the thread which another might need to keep a loom fed. Children, ever eager to take part in adult occupations, would learn to spin at an early age and, when the fly-shuttle made a wide reach for the weaver's arms no longer necessary, would presently be presiding over looms. At how young an age all this started depended upon demands made by parents and on the abilities, mental and physical, of the children. The age of a child thus working would be a secondary consideration, if considered at all. It was all being done in a home workshop, and few people would have dreamed of questioning the prerogatives of parents who might legally claim dominion over their children and over any wages those children might earn until they arrived at the age of majority.

When the work had moved into the large structures where thousands of spindles were turning in huge, cavernous rooms and many, many looms were joining in constant vibration and deafening clatter, the picture of employment, we might think, should have changed. However, neither parents nor children nor employers seemed to have changed from old ways of thought with the change from old-style hand work to new-style machines. Young children continued to work, now walking beside their parents to enter the factories as their parents did, though at a lower wage, which, in any case, was destined to end in the father's pockets.

As the factories grew and home textile manufacturing all but disappeared, there also grew, if slowly and with a limited number of people, an uneasiness as to the wisdom of employing the very young in factories. Baines attempted to deal with this uncomfortable question in an evenhanded manner, trying to explain and justify while revealing the concern he, as well as others, felt about the effect of factory employment on children and the probable consequences to all if the practice were to be interdicted by law.

He described a proposed law which "prohibits the working of

young persons under eighteen years of age for more than twelve hours a day in factories; but as such young persons form nearly one-half of the hands, and are employed in many of the operations, the effect is to limit the labour of adults to the same period." In one district of nearly 68,000 workers, approximately 28,000 were children, 19,000 men, and 21,000 women. No one seemed to have been radical enough to suggest that by not employing children or by shortening their work day from twelve or fourteen hours to a mere ten hours, everyone might have been better off—nobody, that is, save a physician whose suggestions were regarded, even by otherwise high-minded Baines, as "quite impractical."

Baines then evenhandedly quoted "the opinions of a skillful physiologist, the late Mr. Thackrah of Leeds, whose work on 'The Effects of Arts, Trades, and Professions and of Civic States and Habits of Living, on Health and Longevity' displays acute observation and independent thought." Of a mill visited by him in Manchester during the early 1830s, Thackrah's independent thought perceived:

> In this mill 1500 persons are employed, and more than half of these are under the age of fifteen. It is said that none are admitted under that of nine, but several children, from their appearance, we should have supposed a year or two younger. . . . Most of the children are barefoot. The work commences at half past five A.M. and ends at seven P.M., and intervals are allowed of half an hour for breakfast, and one hour for dinner. The mechanics have half an hour also for afternoon meal; but this is not allowed to the children and other operatives. We were informed that at many mills no time is allowed for breakfast, though the work commences as early as half past five. At other mills, moreover, it appears that the dust is much greater, particularly in the carding rooms; and less attention is paid to the health and comfort of the operatives.
>
> I stood in Oxford Road, Manchester, and observed the stream of operatives as they left the mills at twelve o'clock. The children were almost universally ill-looking, small, sickly, barefoot, and ill-clad. Many *appeared* to be no older than seven. The men, generally from sixteen to twenty-four, and none aged, were almost as pallid and thin as the children. The women were most respectable in appearance, but I saw no fresh or fine-looking individuals among them. (quoted by Baines, p. 462)

Those operatives that Thackrah was watching were coming out of spinning mills which, without mechanical ventilation, had lint-filled atmospheres. Weavers, Thackrah believed, were not so badly off as spinning mill operatives, whom he described compassionately: "Here I saw . . . men and women that were not to be aged—children that were never to be healthy adults. . . . I feel convinced that . . . the long confinement in the mills, the want of rest, the shameful reduction of the intervals for meals, and especially the premature working of children, greatly reduce health and vigour, and account for the wretched appearance of the operatives" (quoted by Baines, pp. 462–463).

By 1833, enough people had become concerned to attempt to have a bill passed to regulate the ages of children employed in the mills and the hours they should be expected to work, demanding that they attend school for at least two hours each day six days a

week. In addition, the conditions within the mills themselves were to be improved. One item listed was the whitewashing of the walls at least once a year. Not even mentioned was a provision for adequate ventilation!

Baines, who seems to have been a kindly man in general, admitted:

> Some provisions of this Act have proved to be quite impractica-
> ble. All the Inspectors declare, that the clauses requiring the
> education of the younger children, and forbidding those children
> to be worked more than 48 hours in the week, that is, eight
> hours in the day, have only had the effect of compelling the mas-
> ters to discharge the children between nine and eleven years of
> age. If the Act should continue in force, all children under twelve
> years of age would be discharged in March, 1835, and this
> would make it impossible in many cases to carry on the mills, as
> children above that age could not be had in sufficient numbers.
> The Inspectors, therefore, state that the Act must be amended in
> these respects. . . . It is found impossible to compel the educa-
> tion of the children, and the attempt to do it has only produced
> hardship to them and their parents, from the number who have
> lost their employment. The commissioners had hoped that the
> manufacturers might obtain relays of children, each set working
> no more than eight hours a day, whilst those above 13 years
> worked twelve hours. But neither can the children be obtained,
> nor will the masters submit to the inconvenience caused by the
> change of hands. (Baines, pp. 479–480)

Inconvenience to masters was clearly more to be avoided than damage to workers' health.

"Feeling most sensibly the importance of education to the working classes," Baines continued, "and the undesirableness of working children at a tender age, I am yet convinced that very many of the poor have not the means of educating their children, or of supporting them in idleness; and that, therefore, to forbid the admission of such children into the mills is, in fact, to consign them to the streets and to deprive them of that food which their work might procure." A crucial argument seems to have been this one: "England has manufac-
turing rivals; and if parliament were, from a false humanity, to limit the persevering industry of our workmen, one of our principal advantages over other nations would be sacrificed, and the labourers themselves would be the greatest sufferers."

Was Baines perhaps looking uneasily over his shoulder at the very young textile industry of the young United States, where children were likewise working in mills to add to their parents' income and, as Alexander Hamilton pointed out in his *Report on Manufactures*, thereby to be kept out of the usual mischief of childhood? Baines pointed out:

> Within the last half century cottons to the enormous value of £570,000,000 have been sent from this country to foreign markets. It is obvious that a trade of this magnitude must have contributed largely to sustain that revenue, to prevent the national resources from being intolerably oppressed by taxation and therefore to uphold the power and guard the tranquility of the state. . . .
>
> There are those who prognosticate that she [England] has already reached the highest point and is destined to rapidly decline from it. These individuals apprehend a competition too formidable to withstand, on the part of several foreign nations:—from the United States of America, where the spinning machinery is equal to that of England, where there are thousands of English workmen, where ingenuity and enterprise eminently mark the national character, and where the finest cotton is grown within the States themselves.

He did not raise the question as to what had motivated those English workmen to transfer their activities to America. As an afterthought, apparently, Baines included among possible rivals Belgium and Switzerland as well as "other countries of Europe, where the manufacture exists, and is rapidly extending . . . and the East Indies, where one or two spinning mills have been established and where, in weaving, if not in spinning, the natives are supposed to have a great advantage." In the end, however, Baines concluded that, despite such rivals, England's place "at the head of manufacturing countries" was practically impregnable.

Impregnable it may have seemed, but it was not unassailable. It was being increasingly assailed from across the sea, where the number of English emigrants from Lancashire was suggested by the number of cities and towns bearing names associated with the textile centers of England. There were thirty-seven American Manchesters, ranging from a small village in South Carolina to a great manufacturing city in New Hampshire, fifteen Liverpools, eighteen Prestons, and eight Pendletons. Though some of these places may since have disappeared or been renamed, a substantial number are still in existence.

The first cotton mill in America is said to have been founded in Rhode Island in the year 1790 by one such English emigré, Samuel Slater. Following the custom of Lancashire, Slater would have regarded the employment of children as necessary. In 1827, the Slater and Howard Mill in Massachusetts—possibly an offshoot of the original Slater establishment—was contracting for the labor of one Josiah Mouton together with his six children, the father to be paid $620 for a year's labor by the whole family.

In 1789, a few months before the inauguration of George Washington as first president of the United States, Samuel Slater, English mill worker and mechanic, reached New York. He had managed to escape the general interdiction of emigration of skilled mill workers which had designedly kept all but the most adventurous of his fellows from leaving. Lancashire had no intention of sharing its skills and markets with possible rivals, especially since the American rival had also been a customer for finished cotton piece goods.

Yet perhaps the mill owners of Lancashire were not worrying too much about the emigration of just a few skilled workers, not really able to believe that any other land could set up an industry which could rival their own. Especially they could not have worried too much about the very new nation whose destiny they believed to be purely agricultural—the products of such agriculture being destined for their own land. Or perhaps it was that the mechanics who were or had been involved in British factories had become too numerous to be checked up on, especially once a man had separated himself from mill employment and submerged himself in some other part of the British Isles.

In any case, Samuel Slater managed to get away, taking with him in his head details of the English machines which could not legally be exported either as a whole or as working plans. His arrival turned out to be most opportune for the Quaker firm of Almy and Brown of Providence, Rhode Island. With a long history of successful maritime trade behind the Brown partner, at least, the firm had accumulated enough wealth to be able to afford the risk entailed in starting an almost totally new kind of business for America. It was to be first the spinning into yarn of cotton fiber raised in the South, then converting that yarn into woven fabric.

There was no hypocrisy in the firm's claim of "being desirous of perfecting, if possible, the business of cotton manufacture so as to be useful to the country." Almy and Brown felt strongly that if the young country was to survive in independence and then progress to power, it must look to a future where manufacturing, agriculture, and trade should join in partnership. The spinning machinery then available in America being imperfect, it had become imperative to find someone who knew the Lancashire business and the machines used there. Almy and Brown learned of Slater's presence in America, and by the fall of 1789 they were inviting the Englishman to come to Providence, build a mill, install machinery of the English type (Slater was thoroughly familiar with Arkwright's frame), set the mill in operation, then manage it.

The partners William Almy and Smith Brown were, respectively, the son-in-law and kinsman of Moses Brown, an experienced businessman who was involved in many other projects of his own (White, *Memoir of Samuel Slater*, pp. 72 ff.). The enterprising firm had wide business experience to draw upon, both in the new United States and in older lands across the seas. They had vision enough to see beyond the present struggles of their own land a great and prosperous future, and they had sense enough to choose capable men like Samuel Slater to head their enterprises, then leave the details of organization and management to them.

The spinning frames which Almy and Brown had previously managed to get would not serve, and Slater, immediately aware of their shortcomings, would do nothing with them. He proposed making new ones, using such parts of the old as would answer. Then, with a work force of nine children, Slater was soon turning out satisfactory yarn. Thus, in early 1791, the factory system, with its advantages and disadvantages, was taking it first feeble steps in America.

However, that Rhode Island factory system was yet to include all the steps of cotton preparation from fiber to finished yarn. Part was still being done in homes. Baled cotton came in from the South, purchased there by agents of the factory owners and presumably shipped north in one or another of the Brown family vessels. Once it arrived, the cotton went to various homes where children judged too young to work in factories spread it out, then "picked" and beat it to loosen the fibers and remove specks of dirt—hence the term "cotton batting." Such hand picking was done in homes until about the second decade of the nineteenth century.

The home-picked material that went on to the spinning factories was light and fluffy and ready for carding. Older children spread this on a carding machine, and when it emerged, they put it in a second machine where the cardings were to be formed into loose, soft rolls, called "slivers," ready for the spinning.

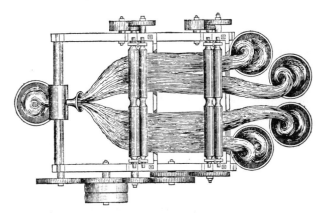

Child labor was, in those days, so generally accepted that even philanthropists like Moses Brown felt no twinges of conscience in employing children, but rather a sense of virtue in making thus available to large, poor families so congenial a means of improving their incomes and status. There were nine children working in the original factory, but by 1801 that number had increased to one hundred, all between the ages of four and ten, working under the supervision of an adult overseer.

There might seem to be some justification in the words Senator Hammond uttered in the course of his famous speech of March 4, 1858: "the Senator from New York said yesterday that the whole world had abolished slavery . . . Aye, the *name*, but not the *thing* . . . The difference between us is, that our slaves are hired for life and well compensated; there is no starvation, no begging, no want of employment among our people, and not too much employment either. Yours are hired by the day, not cared for, and scantily compensated." There were, of course, other differences which he did not mention— that the boasted care and compensation rested upon the whims of the masters, of which Senator Hammond was generally thought to be a good one, and that the slaves could be sold bodily should the master or his creditors so decide.

In the days of the earliest mills, child labor was not an issue, for factory weaving was not to get much of a start in America until after 1809. Soon mill owners were finding that the supply of child workers available locally was running out. They began advertising for large families, preferably those with at least five children. So much the better if these large families were fathered by mechanics or possible overseers. Payment might be in cash or in goods, sometimes overpriced, from the company store. Working hours were from dawn to dusk, the shorter days of winter being somewhat extended by lamp or candlelight.

Today it all sounds pretty grim, but it did not seem so in times when paid employment of any kind was prized by all, even women and children, the menfolk usually pocketing most of the income and few families having surplus to buy themselves even small luxuries. In 1816, the work force of a typical Slater "family" mill consisted of sixty-eight: one family with eight members working; one family with seven members working; two families, each with five members working; four families, each with four members working; five families, each with three members working, plus eight single men and four single women. To Slater, with his background of the Lancashire mills, this type of labor was obvious and highly desirable. People from other backgrounds would presently see it otherwise.

By the year 1906, the system of employment in such mills as Slater's seemed downright inhuman. A memoir published in that year by Stephen Knight, entitled "Reminiscences of Seventy-one Years in the Cotton Spinning Industry" paints this picture:

> On the first day of April in the year of Our Lord one thousand eight hundred and thirty-five, the writer of this paper commenced his labors in a cotton mill as bobbin boy . . . My work was to put in the roving on a pair of mules containing 256 spindles. It required three hands . . . to keep that pair of mules in operation . . .
>
> The running time for that mill, on an average, was about fourteen hours per day. In the summer months we went in as early as we could see, worked about an hour and a half, and then had a half hour for breakfast. At twelve o'clock we had another half hour for dinner, and then we worked until the stars were out.
>
> From September 20 until March 20, we went to work at five o'clock in the morning and came out at eight o'clock at night, having the same hours for meals as in the summer time.

For my services I was allowed forty-two cents per week, which, being analyzed, was seven cents per day, one-half cent per hour.

The proprietor of that mill was accustomed to make a contract with his help on the first day of April, for the coming year. That contract was supposed to be sacred and it was looked upon as a disgrace to ignore the contracts thus made. On one of these anniversaries, a mother with several children suggested to the proprietor that the pay seemed small. The proprietor replied, "You get enough to eat, don't you?" The mother said, "Just enough to keep the wolf from the door." He then remarked, "You get enough clothes to wear, don't you?" To which she answered, "Barely enough to cover our nakedness." "Well," said the proprietor, "We want the rest."

This, perhaps, was his grim idea of a joke, for, as Knight commented, he was, "on the whole as kind and considerate to his help as was any other manufacturer at that time." Knight added, "The opportunities for an education among the factory help were exceedingly limited, as you can well see, both from the standpoint of time and from the standpoint of money." By 1906, when Knight's "Reminiscences" were published, this proprietor would have been looked upon as the most heartless of taskmasters. In the meantime, cotton mills were multiplying all over New England.

15. The Expanding Kingdom

Limits to the number of people locally available as mill employees for profitable mill sites in already settled areas presently began to suggest to would-be mill owners that they should look for new areas in which to erect their mills. They never seemed to consider the possibility of moving south, nearer to where cotton grew. New England still had many untapped sources of water power, and the climate was moist enough to favor the handling of cotton yarns, so the little matter of shipping cotton in from the South was not a worry in an area where seagoing was a flourishing business, often in the families of prospective mill owners.

So they turned their eyes toward the less settled North, starting from Waltham, Massachusetts. The success of the mill there soon suggested other mills on the Merrimack River, where the Massachusetts towns Lowell and Lawrence would be founded as well as the New Hampshire town of Manchester. All these towns on the Merrimack could offer ample water power, though they were near no great falls. However, the large volume of water draining from the White Mountains could be backed up into a great lake and held there until needed for power.

An added advantage was that those rural mountain lands to the north also could supply the much needed workers. But the workers were of a different sort from the chronically poor family groups that came to the mills of southern New England, described by one southern New England employer as "often very ignorant, and too often vicious" (Robinson).

The owners who were bent on founding the new "Waltham system" were determined to guard against the evil of social degradation which they saw in the English system as established in the Slater mills. They would build boarding houses for their employees, keep them under control of the factories, carefully select matrons to supervise the houses, and set strict rules of behavior that must be met by all boarders. With sufficient publicity given to this radically new factory system, which made the places sound more like high-class boarding schools than centers for factory employees, rural parents "were now no longer afraid to trust their daughters in a manufacturing town" (Robinson). In these towns, the girls themselves soon saw to it that undesirables were excluded from the places where they boarded.

Presently it became recognized that apprenticeship in a mill "entailed no degradation of character, and was no impediment to a reputable connection in marriage" (Robinson). Girls arrived from all over rural northern New England, worked for a few years in one textile factory or another, thus helping to pay off the mortgage on the home farm or putting a brother through college or just accumulating enough in her own bank account to help start out on a comfortably debt-free married life.

Victorian though it all may sound to modern ears, the system worked well and the manufacturers who planned it believed that it not only secured a great moral good for the community but, for themselves and their investors, employees of a quality far above that to be found in less carefully organized mills. One employer expressed the belief that with this higher class of employees, he could add 12 percent to 15 percent to the speed of his machines without any increase of damage to the mechanisms. This offered a striking contrast to the English mills, where, if a sufficient labor force was not forthcoming, employers sought extra workers in the poorhouses—workers who had little to offer in skills or in standards of behavior and work.

One of the first northerly towns to be founded under this Waltham system was Lowell, Massachusetts, just south of the New Hampshire line. Standards for employment there were high for the times, as suggested in a report on hours of labor made to the Massachusetts House of Representatives in 1845: "In Lowell, but very few (in some mills none at all) enter into the factories under the age of fifteen . . . Nine tenths of the factory population in Lowell is from the country. They are farmers' daughters. Many of them come over a hundred miles to enter the mills . . . After an absence of a few years, having laid by a few hundred dollars, they depart for their homes, get married and settle in life and become heads of families." Sometimes spoken of as "ladies of the loom," these girls usually lived, while in Lowell, at one of the employer-supervised boarding houses.

These factory girls were generally more literate than average. Some later published sketches of their factory experiences; others developed into campaigners for women's rights. And one such girl was to join the company of Civil War nurses and to record her war experiences in a published book. With the changing times, French Canadians were presently joining the ranks of mill workers, then came European immigrants.

A 1902 visitor from Manchester, England, Thomas M. Young, recorded his impressions of the American mills for the readers of his famous hometown newspaper, *The Manchester Guardian*. Contrasting Lowell with the other Massachusetts mill towns Fall River and New Bedford, he wrote:

> Lowell, an inland town of Massachusetts, owes its cotton mills to the Merrimack River. The oldest and largest of all the mills at Lowell are those of the Merrimack Manufacturing Company. Established in 1822, and rebuilt forty or fifty years ago, they comprise today spinning and weaving mills which contain 144,000 spindles and 4,170 looms, employing 2,300 hands, and printworks which contain twenty-two printing machines and employ 1,000 hands . . .
>
> The mills at Lowell—brick buildings six stories high—are driven partly by water and partly by steam. The water power is obtained from a canal, shaded on both sides with old trees, and flowing soft and clear down the middle of the street leading to the mills and is an ornament as well as a profitable servant to the town.

It would, of course, be the New Hampshire Manchester that roused the greatest interest in the visitor from the English Manchester—also the greatest admiration: "None of the manufacturing towns of New England pleased me so much as Manchester, New Hampshire. Unlike its great godmother, it has clear air, clear waters, and sunny skies; almost every street is an avenue of noble trees, whose leaves fall so thick in autumn on the electric car tracks that at first, when they are sappy, they make the car wheels skid, and later, when they are dry, they are fired by the sparkling current, and fill the city with aromatic smoke."

After paying tribute to the many well-spaced parks, Young gives his impressions of the mill buildings themselves:

> Perhaps the handsomest, certainly the most impressive buildings in Manchester are the Amoskeag and Manchester Mills. They are not ornate—ornate mills are often hideous—but they are built of a warm red brick, beautifully weathered, and form a continuous curved facade (like the concave side of Regent Street in London) half a mile long. Rising sheer out of a deep, clear, swift-flowing stream (the Merrimack), upon the other bank of which are grass and trees, they need little more than be silent to masquerade successfully as ancient colleges.
>
> Not until one has passed over one of the pretty bridges and penetrated through the waterside buildings to the court beyond does one appreciate the enormous extent of these simple, stately buildings. Behind the riverside pile there runs a courtyard so long as to be more like a private road, and on the other side of this road, runs another line of mills, parallel with the curve of the first, so that one cannot see to the end of them.

Praise, indeed, from a visitor from New Hampshire's "great god-mother" and, of course, commercial rival. The visitor continued his tour at the less impressive mills in Maine, then turned south to visit the burgeoning mills of that area, where he found conditions dismal. Yet despite those conditions, the southern mills were already overtaking the northern, as the subsequent decades were to reveal.

It was to be a long time before southern mills would be troubled by investigations of committees on hours of labor, as the northern ones had been in 1845. The investigators were impressed then with how demanding mill work could be. It would look even worse today when an eight-hour day, or forty-hour week, is thought of as too demanding. Yet to the very youngest workers, it did not seem altogether grim. In 1898 was published a book of reminiscences—*Loom and Spindle: Or, Life among the Early Mill Girls*—written by Harriet Robinson, who had been one of those young workers. Harriet's mother had been left widowed in 1831 with four young children, the eldest only seven years old. She had to feed and clothe them on the meager income she could get from a boarding house for mill workers.

Harriet reminisced:

I had been to school constantly until I was about ten years of age when my mother, feeling obliged to have help in her work besides what I could give, and also needing the money I could earn, allowed me, at my urgent request (for I wanted to earn *money* like the other little girls), to go to work in the mill. I worked first in the spinning room as "doffer" [the very youngest children whose task it was to doff, or take off the full bobbins and replace them with empty ones].

I can see myself now racing down the alley, between spinning frames, carrying in front of me a bobbin-box bigger than I was. These mites had to be very swift in their movements, so as not to keep the spinning-frames stopped long, and they worked only

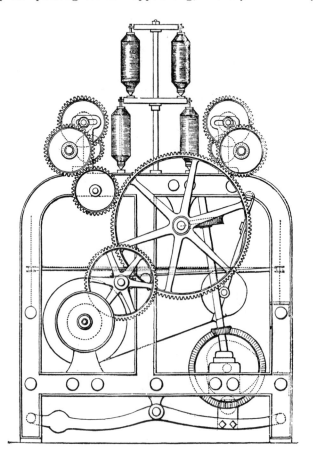

about fifteen minutes in every hour. The rest of the time was their own, and when the overseer was kind they were allowed to read, knit, or even go outside in the mill-yard to play.

Some of us learned to embroider in crewels, and I still have a lamb worked on cloth, a relic of those early days when I was first taught to improve my time in the good old New England fashion. When not doffing, we were often allowed to go home for a time, and thus we were able to help our mothers in their housework. We were paid two dollars a week; and how proud I was when my turn came to stand on the bobbin-box, and write my name in the paymaster's book, and how indignant I was when he asked me if I could "write." "Of course I can," said I, and he smiled as he looked down at me.

The working hours of all girls extended from five o'clock in the morning until seven in the evening, with one half hour for breakfast and for dinner. Even the doffers were forced to be on duty nearly fourteen hours a day, and this was the greatest hardship in the lives of these children. For it was not until 1842 that the hours of labor for children under twelve were limited to ten per day; but the "ten-hour law" was not passed until long after some of these little doffers were old enough to appear before the legislative committee on the subject and plead, by their presence, for a reduction of the hours of labor.

I do not recall any particular hardship connected with this life, except getting up so early in the morning, and to this habit, I never was, and never shall be, reconciled, for it has taken nearly a lifetime for me to make up the sleep I lost at that early age. But in every other respect it was a pleasant life. We were not hurried any more than was for our good, and no more work was required of us than we were able easily to do.

Most of us children lived at home and we were well fed . . . eating substantial meals (besides luncheons) three times a day. We had very happy hours with the older girls, many of whom treated us like babies, or talked in a motherly way . . . and in the long winter evenings, when we could not run home between doffings, we gathered in groups and told each other stories, and sung the old-time songs our mothers had sung . . .

And we told each other of our little hopes and desires, and what we meant to do when we grew up . . .

Holidays came when repairs to the great mill-wheel were going on, or some spring freshet caused the shutting down of the mill; these were well improved. With what freedom we enjoyed those happy times!

All this time Harriet's mother was carrying on the boarding house for her family's support, with as much help from her children as she could get. Of this Harriet wrote:

> We children were expected to scour all the knives and forks used by the forty men-boarders, and my brothers often bought them-selves off by giving me a trifle, and I was left alone to do the whole . . . I don't know why I did not think such long tasks a burden, nor of my work in the mill as drudgery. Perhaps it was because I *expected* to do my part towards helping my mother to get our living, and had never heard her complain of the hard-ships of her life. . . .
>
> I was a "little doffer" until I became old enough to earn more money; then I tended a spinning-frame for a little while; and after that I learned, on the Merrimack Corporation, to be drawing-in girl, which was considered one of the most desirable employments, as about only a dozen girls were needed in each mill. We drew in, one by one, the threads of the warp, through the harness and reed, and so made the beams ready for the weaver's loom. I still have the two hooks I used so long, compan-ions of many a dreaming hour as the "badge of all my tribe" in drawing-in girls . . .
>
> When I look back into the factory life of fifty or sixty years ago, I do not see what is called "a class" of young men and women going to and from their daily work like so many ants that cannot be distinguished one from another; I see them as individuals, with personalities of their own . . . Yet they were a class of fac-tory operatives and were spoken of (as the same class is spoken of now) as a set of persons who earned their daily bread, whose condition was fixed, and who must continue to spin and weave to the end of the natural existence . . . It was the good fortune of these early mill-girls to teach the people of that time that this sort of labor is not degrading; that the operative is not only "ca-pable of virtue," but also capable of self-cultivation.

This mill girl, gifted writer of the above memoirs, lived until 1911, to the age of ninety-six. She became a fighter for women's rights and one of the organizers of the General Federation of Women's

Clubs. Another Lowell mill girl was the schoolteacher and author Lucy Larcom, who attained some modest renown as a poet. She did not work in the mills as long as Harriet Robinson, but did remain there long enough to view the great mill wheel at work in the basement and report what she saw with imaginative eyes.

Clearly, those who desired an education, even though committed to mill work, somehow managed to get it. Harriet Robinson reports:

> Some of us were fond of reading and we read all the books we could borrow. One of my mother's boarders, a farmer's daughter from "the State of Maine," had come to Lowell to work for the express purpose of getting books, usually novels, to read, that she could not find in her native place. She read from two to four volumes a week; and we children used to get them from the circulating library, and return them, for her. In exchange for this, she allowed us to read her books, while she was at work in the mill; and what a scurrying there used to be to be home from school, to get the first chance at the new book! In this way, I read the novels of Richardson, Madame D'Arblay, Fielding, Smollett, Cooper, Captain Marryatt, and many another old book.

Quite a solid diet for a very young girl!

Like Harriet Robinson and also the daughter of a widowed mother, this one with eight children to feed and clothe, Lucy Larcom shouldered her share of family responsibilities by going into the mills. In her book *A New England Girlhood*, she wrote of this experience:

> The older members of the family did everything they could, but it was not enough. I heard it said one day, in a distressed tone, "The children will have to leave school and go into the mill."
>
> There were many pros and cons between my mother and sisters before this was positively decided. The mill-agent did not want to take us two little girls, but consented on condition we should be sure to attend school the full number of months prescribed each [at that time, three]. I, the younger one, was between eleven and twelve years old . . .
>
> Once, before we left our old home, I had heard a neighbor condoling with my mother because there were so many of us, and her emphatic reply had been a great relief to my mind: "There isn't one more than I want. I could not spare a single one of my children."

But her difficulties were increasing, and I thought it would be a pleasure to feel that I was not a trouble or burden or expense to anybody. So I went to my first day's work in the mill with a light heart. The novelty of it made it seem easy and it really was not hard, just to change the bobbins on the spinning-frames every three quarters of an hour or so, with half a dozen other little girls who were doing the same thing . . . There were compensations for being shut in to daily toil so early. The mill itself had lessons for us. But it was not, and could not be, the right sort of a life for a child.

Harriet Robinson, on the other hand, saw employment in the mills rather differently: "As the cotton factory was the means of the early schooling of so large a number of men and women, who, without the opportunity thus afforded, could not have been mentally so well developed, I love to call it their *Alma Mater* . . . The discipline the work brought us was of great value . . . we worked and played at regular intervals, and thus our hands became deft, our fingers nimble, and we were taught habits of regularity and industry."

Such fond remembrances may have been no more than the nostalgia of a woman well past her seventieth birthday looking back wistfully on her own youth. Yet her highly literate expressions reveal penetrating habits of thought and practice in expressing such thoughts. For some of this, at least, credit should be given to the short-lived *Lowell Offering*, a magazine organized, written, edited, and published by mill girls of the 1840s and with sufficient competence to astonish visitors from England like Anthony Trollope, Harriet Martineau, and Charles Dickens.

To Dickens' eyes, Lowell looked young and fresh (it then being scarcely two decades old) except for the mud "which in some parts was almost knee-deep, and might have been deposited there on the subsiding of the deluge . . . There are several factories in Lowell . . . I went over several of these . . . and saw them in their ordinary working aspect, with no preparation of any kind or departure from their everyday proceedings. I may say that I am well acquainted with our manufacturing towns in England, and have visited many mills in Manchester and elsewhere in the same manner."

What particularly impressed Dickens, who in general had little good to say of things American, was the workforce—clean, apparently healthy, and well dressed. He also commended their surroundings: "The rooms in which they worked were as well ordered as themselves. In the windows of some there were green plants, which were trained to shade the glass; in all there was as much fresh air, cleanliness and comfort, as the nature of the occupation would possibly admit of."

Dickens described with praise the boarding house residence, the limitations set on child labor (no more than nine months in the year), the infirmary and hospital. But to us of nearly a century and a half later, his most interesting comment is embodied in a little paragraph clearly aimed at English readers:

> I am now going to state three facts, which will startle a large class of readers on this side of the Atlantic very much. Firstly, there is a joint stock piano in a great many of the boarding houses. Secondly, nearly all these young ladies subscribe to circulating libraries. Thirdly, they have got up among themselves a periodical called *The Lowell Offering*, a repository of original articles, written exclusively by females actively employed in the mills,— which is duly printed, published and sold, and whereof I brought away from Lowell four hundred good solid pages, which I have read from beginning to end.

Dickens knew that English readers, with the mill towns of Lancashire in mind, would regard an undertaking like the mill girls' *Lowell Offering* preposterously "above their station." The social reformer in Charles Dickens summed up his own reaction with: "I know of no station which is rendered more endurable to the person in it, or more safe to the person out of it, by having ignorance as an associate. I know of no station which has the right to monopolize the means of mutual instruction, improvement, and rational entertainment."

Of the contents of the *Offering* itself, Dickens wrote: "I will only observe, putting entirely out of sight the fact of the articles having been written by these girls after the arduous labours of the day, that it will compare advantageously with a great many English Annuals. It is pleasant to find that many of the Tales are of the Mills and those who work in them."

It may have come as something of a surprise even to Dickens to find in the *Offering* one writer's shrewd analysis of the motives that had brought girls from their rural homes to work in the mills: "There are girls here for every reason, and for no reason at all. I will speak to you of my acquaintances in the family here. One who sits at my right hand at the table, is in the factory because she hates her mother-in-law" (Eisler). Here read "stepmother" and visualize a latter-day Cinderella. The writer goes on down the line of diners seated at her boarding house table, listing penurious or over-pious parents, until she reaches the one who "is here because she must labor somewhere, and she had been ill treated in so many families that she has a horror of domestic service. The next has left a good home because her lover, who has gone on a whaling voyage, wishes to be married when he returns, and she would like more money than her father will give her. The next is here because her home is a lonesome country village, and she cannot bear to remain where life is so dull. The next is here because her parents are poor, and she wishes to acquire means to educate herself . . ." (Eisler).

The fascinating revelation in this analysis of motives is not that the girls loved mill work as work—some *Offering* writers were openly critical of the mills—but that, for such girls, the mills' chief function was to serve as a means to other, more personal ends, a function which would not last when conditions of employment changed. The near-idyllic situation, as a number of the earliest mill girls seemed to think of it, was, in fact, already beginning to weaken even as the *Lowell Offering* first appeared on the scene. Students of labor movements can explain this situation with thousands of reasons, few of which can find a place in a book primarily concerned with cotton. But already in the 1840s, increasing competition between textile mills and gradually widening opportunities for women in other areas of employment began to play their part.

Under the growing pressure of competition, wages and working conditions began to decline. Machines were speeded up, and the kind of girl who had once eagerly sought employment in the mills no longer found the mills of Lowell or Lawrence or Manchester highly desirable places to work. They found other means of accumulating bankrolls, and, in any case, just stopped looking for places in the mills. Their places were taken by increasing numbers of immigrants, generally less well educated, less committed to the mills, glad at first to find work at wages of almost any kind. From all these causes stemmed the labor troubles, descending occasionally to downright vandalism, that in the nineteenth century and on into the twentieth, haunted the New England textile industry and sent it scurrying south.

Yet, strangely, elderly former mill workers who survived into the mid-twentieth century seemed to recall mills like the Amoskeag of Manchester, New Hampshire, with a sort of nostalgia. Not glossing over the fact that conditions were demanding, hours long, and pay small, there seems to run through their reminiscences a sort of regret. Many of their names suggest French Canadian origins, their recorded memories generally including large families, poverty, and the need for enhanced family income. Yet one who reluctantly started mill work as a teenager admitted that if she had to start over again, she wouldn't mind working in a mill "if it was like the Amoskeag." What the workers who did not survive so long might have recalled of their days in the mills we cannot know. Certainly we know, as they could not, that behind it all was King Cotton's drive for expanding dominion.

16. The King's Courtiers

Though the cotton plant itself could not escape involvement in almost every aspect of the textile industry, it can hardly be held totally responsible for the changing conditions in Northern textile mills before the 1860s, when the Civil War, whose root cause was cotton itself, broke shatteringly upon the national scene. For the South, where cotton had grown to be the major crop, cotton had to assume, directly or indirectly, a large share of the blame—almost as much as Senator Hammond had proudly assigned to it.

If social conditions in the mid-nineteenth-century mill towns of England and New England left something to be desired, they left still more on the Southern plantations where cotton was grown and harvested. Plantation owners, of course, were rarely aware of this or, if vaguely aware, would not publicly confess to such a heresy. Their lives, in general, went along smoothly and happily in great houses where good books, good music, good food, and good liquor held sway. The nitty-gritty of everyday life was relegated to black-skinned menials whose own lives were assumed by most planters to belong in a condition where no alteration was needed, who happily accepted their own servitude and wisely avoided thinking about changing their status.

In an earlier part of his already quoted cotton-enthroning speech of March, 1858, Senator Hammond had put that point of view in a nutshell:

> In all social systems there must be a class to do the mean duties, to perform the drudgeries of life; that is, a class requiring but a low order of intellect and but little skill. Its requisites are vigor, docility, fidelity. Such a class you must have or you would not have that other class which leads progress, refinement, civiliza-

tion . . . Fortunately for the South, she found a race adapted to that purpose to her hand—a race inferior to herself, but eminently qualified in temper, in vigor, in docility, in capacity to stand the climate, to answer all her purposes. We use them for the purpose and call them slaves . . . I will not characterize that class at the north with that term; but you have it; it is everywhere; it is eternal.

Thus spoke the self-deluded vassal of King Cotton, the feudal lord who held sway over hundreds of serfs who could not leave his plantations unless he were to sell them bodily. Yet the progress, refinement, and civilization which Senator Hammond and his fellow Southerners believed in, as they believed in slavery, were not to prove as eternal as he claimed. Three years and one month later, those beliefs began, by the South's own choosing, to face the crucial test.

Not all Southerners had been equally blind to the risks inherent in their chosen way of life. Already in 1845, a southern writer named William Gregg, representing a tiny minority, was forecasting disaster: "Since the discovery that cotton would mature in South-Carolina, she has reaped a golden harvest; but it is feared that it has proved a curse rather than a blessing, and I believe that she would at this date be in a far better condition, had the discovery never been made" (quoted by Mitchell, *The Rise of Cotton Mills in the South*, p. 28). Another Southerner, D. A. Tompkins, expressed a similar opinion during the post-war decades: "The invention of the cotton gin . . . before 1860 . . . was nearer anything else than a blessing. It was primarily responsible for the system of slavery. . . . Cotton . . . in its manufacture . . . is the life of the South, but we could probably have done as well without it until we began to manufacture it" (ibid.). Respect for manufacturing was something the Southern plantation owners learned slowly.

As the South presently came to realize, the region would certainly have done far better without the planting of cotton, which was the keystone of that structure of elegance, refinement, and civilization of which Senator Hammond had boasted. Yet, insofar as the institution of slavery was concerned, there were many other Southern dissenters. Emotionally arrayed against slavery were many women who lived with the unhappy conviction that concubinage, which had grown apace since the constitutional interdiction of the importation of slaves from Africa went into effect in 1808, was a threat to the stability of their lives. Breeding slaves had become a business that a plantation owner and his sons did not hesitate to indulge in.

In March, 1861, Mary Chestnut, that brilliant South Carolinian whose Civil War diary was to be published a half century later, put the women's feelings into moving words: "I wonder if it be a sin to think slavery a curse to any land. Sumner said not a word of this hated institution which is not true. Men and women are punished when their masters and mistresses are brutes and not when they do wrong—and then we live surrounded by prostitutes . . . God forgive us, but ours is a *monstrous* system and wrong and iniquity. Perhaps the rest of the world is as bad—this *only* I see."

This she had written before Fort Sumter was fired on and general hostilities broke out. But as months passed and as the war, which she and her coterie had believed must be ended within a few months with the South triumphant, dragged on, the sentiments she recorded in her diary were less clear and uncompromising. She never supported the institution of slavery, but her contempt for Yankees, including the Senator Sumner whose condemnation of slavery she had previously endorsed, grew. The brutal masters and mistresses were relegated to a minority, and she rather assumed the contented acceptance of their lot by slaves while she wondered uneasily what kind of thoughts might be running through the brains behind those dark and smiling faces. Most of all, her anti-Yankee sentiments grew more and more explicit. She referred to those New Englanders living "in nice New England homes—clean, clear, sweet-smelling—shut up in libraries, writing books which ease their hearts of their bitterness to us, or editing newspapers, all of which pays better than anything else in the world. Even the politician's hobbyhorse—antislavery—is the beast to carry him highest."

Senator Sumner, as Mary Chestnut could not escape knowing, had not been living at ease. He had dedicated his life to the cause of abolition of slavery within the Union, delivering in the Senate long and vituperative speeches. In one of these he had singled out with special venom a Senator Butler of South Carolina. Preston Brooks, a congressman from the same state and a relative of Senator Butler, considered himself honor-bound to "punish" the outspoken Massachusetts senator. He sought out the latter while he was bent over his desk in the Senate chamber, and, without warning, beat him into insensibility. It was a near fatal attack from which the New Englander took months to recover.

Such was the violence, personal and regional, which was then brewing and which Mary Chestnut's South, to maintain its valued way of life, endorsed to the point of seceding from a Union upon which so many great and revered Southerners had, less than a century before, staked life, liberty, and sacred honor. Such were the tyrannical and subversive powers of King Cotton.

In the 1850s, the cultivation of cotton was going on as it had gone on for decades and as plantation owners like senators Hammond and Butler expected it to go on indefinitely. After the wide fields were cleared of refuse and old stalks during January and February, they were ploughed and planted in late March or early April, a girl with her apron full of seed following the plow to drop seed in its wake. Six to ten days later, small green shoots would appear and cultivation began, experienced field hands using their hoes to encourage the best shoots and discourage others. *Harper's Magazine* in 1854 described this process of "leaving the incipient stalk unharmed and alone in its glory; at nightfall you can look along the extending rows, and find the plants correct in line, and of the required distance of separation from each other." Clearly, this required both skill and intelligence in the workers, Senator Hammond's strictures about those qualities notwithstanding. For planter and overseer, such a field would be a matter of pride, though both knew that "the vicissitudes attending the cultivation of the crop have only commenced." Weeds, drouth, and an almost infinite number of plant pests waited, ready to take their toll.

The progress of cotton from planted seed to matured fruit has rarely been unimpeded. By July, flowering may have begun. "The cotton blossom is exceedingly delicate in organization. It is, if it is perfect . . . of a beautiful cream color. It unfolds in the night, remains in its glory through the morn—at meridian it has begun to decay. The day following its birth it has changed to a deep red, and, ere the sun goes down, its petals have fallen to the earth, leaving inclosed in the capacious calyx, a scarcely perceptible germ. This germ, in its incipient and early stages is called 'a form,' in its more perfected state, 'a boll.'" It is these bolls that eventually burst open to scatter "their rich contents to the ripening winds"—unless, of course, those contents have promptly been gathered in.

Rust and rot of leaves or bolls, as well as worms, weevils, and such threaten yields. The wonder is that King Cotton's throne so long remained all but impregnable despite the general early ignorance of disease-causing organisms and of ways to control such diseases as well as the worms and insects that fed upon cotton plants. In this ignorance, huge areas were planted to the single crop, then replanted year after year, thus inviting and nourishing enemies which, in the wild, would have found it less convenient to move from one cotton plant to another. As with potatoes in the Ireland of the 1840s, a crop thus cultivated could be headed for extinction. Thus, even before war devastated the land, King Cotton's status as a uniquely powerful monarch was hastening toward an end.

The self-deluded planters, however, continued to believe in King Cotton's invincibility, to worship at his throne, and to cultivate cotton on the same acreage using the same routines year after year. They neglected either to seek out new lands or find ways to restore fertility to the old depleted ones. When they finally learned to mend their ways, cotton production again began to climb.

From late July until about Christmas, the fiber was gathered in from the opened bolls, each field hand gathering from 250 to 300 pounds each day. At evening, the bags of harvested cotton were delivered to the second floor of the gin house and dumped into a hopper that led to the gin on the floor below. Ginning cotton was a crucial step.

As described in *Harper's Magazine*, "Much of the comparative value of the staple of cotton depends upon the excellence of the cotton-gin. Some gins separate the staple from the seed far better than others, while all are dependent more or less for their excellence upon the judicious manner they are used. With constant attention, a gin-stand impelled by four mules, will work out four bales of four hundred and fifty pounds each day . . . Upon large plantations the steam engine is brought into requisition, which, carrying any number of gins required, will turn out the necessary number of bales per day." Clearly, the mule was facing extinction as a powerer of engines.

The annual cotton ritual came to an end with baling, accomplished by "a single but powerful screw . . . When a sufficient quantity has been forced by 'foot labor' into the press, the upper door shut down, and the screw is applied, worked by a horse." Then the bagging, previously set in place, was drawn up around the bale and sewed up with twine. And so, while the bales went on their way via warehouse, rail, boat, and, preponderantly still, transatlantic steamer, the cotton year came to an end in time for a brief Christmas holiday before fields were cleared and prepared for yet another spring planting.

17. King Cotton Seeks Allies

Some baled cotton headed north to New England, but most was destined for the English port of Liverpool and thence to Lancashire, whose mills, and of course people, had come to be increasingly and irreversibly involved with cotton from the American South. "For nearly a century," Sir Arthur Arnold wrote of the Lancashire of 1864, "she has been offering the working classes such reward for their labours as it was impossible to obtain elsewhere." The word got around, as such words are bound to do, and the working classes flocked to Lancashire from all over England and from Ireland, too, swelling the population of the area. Lancashire, about four percent of the area of England, soon contained about twelve percent of the country's total population—about two and a half million persons having become dependent, in one way or another, upon King Cotton and on the industry he sponsored.

Few of those British workers could have given much thought to where the cotton grew, how it was cultivated, or under what conditions the cultivators had to work. It was a remote world both in distance and in the kinds of people involved there. It was not their concern—not until it was unavoidably thrust upon their consciousness by events over which neither they, their employers, nor their government seemed to have any control.

Yet some of the workers must have been growing uneasily aware of the menacing clouds gathering in the west, while those who could read began to watch the news with growing concern. On January 24, 1861, a young worker who was keeping a personal diary noted, "We have got a notice put up in our mill today, giving us notice of a reduction in wages. It is upon account of the disturbance in America. The cotton market has risen in speculation that there will be no crop next year if the civil war should happen in the United States" (cited by Longmate). It happened, of course, before three months had passed and at Fort Sumter in South Carolina, native state of the ardent secessionist, Senator James Henry Hammond.

The fears of the young diarist and of his coworkers were soon being realized. By late 1861, a notice was being put up in a Bolton mill: "No cotton—no work," summarizing succinctly the prospect for all the cotton mills of Lancashire. The prospect, of course, was for disaster, as a visitor to Preston would record: "I was astonished by the dismal succession of destitute homes, and the number of struggling owners of little shops, who were watching their stocks sink gradually down to nothing and looking despondingly at the approach of pauperism. I was astonished at the strings of dwellings, side by side, stripped, more or less, of the commonest household utensils—the poor little bare houses, often crowded with lodgers, whose homes had been broken up elsewhere; three or four families of decent working people sleeping anywhere, on benches or straw; and afraid to doff their clothes at night because they had no other covering" (cited by Longmate).

A woman, coming from London to help a harassed local clergyman in his attempts to care for the newly poor, recorded her impressions in December, 1862:

It is not the operatives alone who are suffering from the crisis; the entire human machinery in each district is at a standstill. Clerks, shopkeepers, mechanics, warehousemen, tradesmen of every grade—all whose business depended on the operatives, are also involved in the same ruin; and despair and perplexity are written on the countenance of these once prosperous people. No words can possibly estimate the destitute condition which eighteen months of forced idleness, and the almost entire stagnation of business, has produced amongst them.

Such an overwhelming calamity . . . could neither have been contemplated nor forestalled. The blast of desolation has swept over the entire land, levelling all alike . . . Step by step, want has overtaken them like an army cut off from its supplies; and the charity of the world now alone stands between these 500,000 men and starvation. (cited by Longmate)

Presently, more efficiently organized relief measures were to help make the subsequent winters of the cotton famine less totally calamitous. And all this misery was due to a plant few Lancashiremen, workers or factory owners, would ever meet in its native habitat. Yet so preeminent a place did cotton hold in the minds of Lancashiremen that the *Times* of London was moved to quip in January, 1861, "No Manchester jury was ever found to acquit the unhappy wretch who was accused of stealing cotton" (cited by Longmate).

Of course, British sympathies would at first be much inclined toward the cotton-raising South, whatever the feelings on the subject of slavery in general which, many Britons were proud to reflect, had been abolished throughout the empire in 1833. Whether the destitute factory workers shared that pride altogether is doubtful, if they were in any condition to take pride in anything that could not mitigate their miseries. It is also doubtful that the sufferers would have rejected any chance for employment, even if the price to be paid for such was the entrenchment of slavery in a faraway America none of them ever expected to see. As for their late employers, empty factories and emptying pocketbooks spoke louder than ideas.

Besides, during the early decades of the nineteenth century, Liverpool, through which most raw cotton was imported, had hosted more than a hundred vessels engaged primarily in the slave trade. In 1833, when the British Empire rejected slavery in all its parts, those vessels had quickly changed their destination to New Orleans, where they might fill their holds with raw cotton destined, via Liverpool, for the factories of Lancashire. The Cotton Exchange of Liverpool was not only the heart of the city's business but also its most exclusive club. Thus, self-interest, along with a conviction that what was happening on the far shore of the Atlantic represented a threat to property as well as to their cherished way of life, long prompted importers of Liverpool and factory owners of Lancashire to look on the American South with eyes at once sympathetic and covetous.

There were, however, many in England whose sympathies were with the North, and these promptly found a spokesman in John Bright, a member of Parliament. Bright was a friend and correspondent of Charles Sumner, the famous American abolitionist senator who in 1856 had, when on the floor of the Senate, been physically attacked and seriously injured by a colleague from the South. As time passed, increasing numbers of Englishmen began to listen to Bright, notably after President Lincoln's announcement, on September 22, 1862, that after January 1, 1863, all slaves in the seceding states were to be considered free. Some propertied Britons, however, were apprehensive lest this act spark slave uprisings all over the world. Yet an increasing number of Britons began to express explicitly pro-North sympathies.

In December of 1862, John Bright wrote to Sumner: "The anti-slavery sentiment here has been called forth . . . since the Proclamation was issued and . . . every day the supporters of the South among us find themselves in greater difficulty owing to the course taken by your government." On the eve of the day the Emancipation Proclamation was to go into effect, there was a meeting "composed mainly of the industrial classes of Manchester," attended by an enthusiastic audience which announced profound sympathy with "the efforts of President Lincoln and his colleagues to maintain the American Union in its integrity" and, of course, to free the slaves. Within a month, a similarly crowded pro-North rally was being held in London, so that John Bright could report to his American correspondent, "Our Southern newspapers are surprized and puzzled at the expression of opinion in favor of the North" (Bright, *American Question*, pp. 181, 182, 186).

The pro-South news sheet, *Index*, had an answer for this, if a revealing one. The British authorities, it averred, were ready to "let Lancashire starve . . . they will submit to a blockade of Bermuda or of Liverpool; but they will do nothing which may tend to bring a supply of cotton from the South" (cited by Longmate, p. 267)—excepting, of course, those praiseworthy businessmen who had taken up a "cotton loan" for building armed ships for the Confederacy.

In March, 1863, a pro-North rally was called at St. James Hall, London, to which—presumably at the suggestion of Karl Marx—trade unions sent large delegations. John Bright was at his most eloquent:

> The distress in Lancashire comes from a failure in the supply of cotton; that comes from the war in the United States; and that war has originated from the efforts of slaveholders in that country to break up what they themselves admit to be the freest and best government that ever existed, for the sole purpose of perpetuating the institution of negro slavery. . . . There may be men . . . sitting among your legislators who will build and equip . . . ships to prey upon the commerce of friendly powers, who will dis-

regard the laws and honour of their country. . . . There may be
men, too, rich men in this city of London, who for the chance of
more gain than honest dealings will afford them would not hesitate to assist a cause whose fundamental institution is declared to
be . . . infamous by the statutes of their country. I speak not to
these men, I leave them to their consciences. (Brigno, *Speeches*,
pp. 127, 128)

Bright's speech met with deafening applause, to the delight of young
Henry Adams, grandson of presidents and private secretary to his
father, Charles Francis Adams, then United States ambassador to
England.

Young Henry Adams had personal as well as patriotic reasons for
feeling delighted with the loud expressions of Union sympathy. He
was painfully aware that, for the English landed gentry, the South
and its way of life—great houses set amidst wide estates cultivated
by a docile servant caste—seemed far more congenial than did the
industrial North, which even cultured nineteenth-century Englishmen like Charles Dickens and Anthony Trollope generally looked
upon with negative enthusiasm. As the prominent Englishman
Leslie Stephen put it, succinctly explaining the reaction of his own
class, it was "a time . . . when the North had scant justice and no
mercy at their hands. I myself have seen that most distinguished
man, Charles Francis Adams, subjected in society to treatment
which, if he had resented it, it might have seriously imperilled the
relations of the two countries." Stifling all personal feelings, the se-
nior Adams treated the snubs, social as well as professional, with
"mute disdain," thus helping the North's sympathizers in Parliament
keep England neutral and preventing it from making "the ruinous
mistake of taking part with the South."

That neutrality was, for long, highly questionable as England tee-
tered on the brink of making that ruinous mistake. As Foreign Minister Lord John Russell admitted to Charles Francis Adams, British
merchants "would, if money were to be made by it, send supplies
even to hell at the risk of burning their sails." They were already doing approximately that in outfitting and supporting blockade runners to Southern ports. They were doing yet worse, as witnessed by
the American consul at Liverpool who was keeping an eye on the
ships being constructed on Merseyside. Soon he was recognizing the
hull anonymously listed as "Number 290" as a fast and powerful

commerce destroyer. She would presently sail under British crew and officers to serve the Confederacy as the *Alabama*—a blatant violation of neutral status which was, for decades to come, to bring blushes of shame to the cheeks of many Britons.

Even English apologists for the North were not being given an easy time in England. Journalist and clergyman Leslie Stephen wrote his mother in 1864 from Washington, where he would gladly have been able to act as apologist for his own land, had circumstances justified it:

> I really don't know how to translate into civil language what I have heard a thousand times over in England; that both sides are such a set of snobs and blackguards that we only wish they could both be licked; or that their armies are the scum of the earth and the war got up by contractors; or that the race is altogether degenerate and demoralized, and it is really pleasant to see such a set of bullies have a fall . . . I really can't tell them all these little compliments, which I have heard in private conversation, word for word, and which are a free translation of *Times* and *Saturday Review*, even if I introduce them with the apology . . . that we know nothing at all about them.

In England, agents from the South and their publication *Index* were doing their best to spread the truth "about them" as they wished Britons to see it. It was, perhaps, inspired by a growing desperation, but the wind was veering against them as the enthusiastic popular reception of John Bright's speech had demonstrated. So, with pro-North sentiments thus enthusiastically and publicly expressed and with consciences of more restrained Englishmen becoming aroused, England was to refrain from officially espousing the cause of the Confederacy.

The Confederates, however, clung to the idea and the hope that some foreign nation should declare for it. In July, 1861, Mary Chestnut wrote rather wistfully, "How we cling to the idea of an alliance with England or France! Without France, even Washington could not have done it." A few weeks later, when the Confederacy was beginning to feel the first slight pinch of the blockade, she recorded, "The Lord help us since England and France won't or don't. If we could only get a friend outside and open a port!"

What Mary Chestnut and her circle of the elite of the Confederacy would not or could not recognize in time was that their South, long prospering on the yield of a predominantly single-crop agriculture, had let herself become tragically dependent upon that crop and upon the goods the sale of that crop could bring her from distant markets. So while Chestnut had to admit that England was remaining "sturdily neutral"—as far as declaring for either side was concerned—the ports of the South continued to be blockaded: "ruling the waves does not help Britannia bring her wares in."

A year later, Chestnut's humorous note weakened: "Can we hold out if Britain and France hold off?" Still two years later, she was sensing the answer as she reported some of the endless discussions that were going on among highly placed Confederates: "Now England and France are not mentioned—and once we counted so strongly on them for a good stout backing. We thought that this was to be a bloodless duel and we would get out of the Union—because they hated us—and why, then, should they want to keep us?"

"Oh, for a single port!" she cried later in the same day's entry. "If the *Alabama* in the whole world had had a port to take her prizes—refit—&c&c, I believe she would have borne us through—one single point by which to get at the outside world and *refit* the whole Confederacy."

To Prussian army officer General Friedrich von Bernhardi, writing a half-century later "out of the fullness of my Germanic heart" and with World War I hopefully in mind, here was an opportunity Englishmen had been too obtuse to exploit as they might: "Since England committed an unpardonable blunder, from her point of view, of not supporting the Southern States in the American War of Secession, a rival to England's world-wide Empire has appeared on the other side of the Atlantic in the form of the United States of North America, which are a grave menace to England's fortunes. The keenest competition now exists between the two countries"—a competition which was causing the German general to lick his lips in anticipation of the time, soon to come, he hoped, when his own country might knock the two great rivals off, one by one.

The general went on to underline "the supineness for which England has herself to blame, when she refused assistance to the Southern States in the American War of Secession, and thus allowed a power to rise in the form of the United States of North America which already, although barely fifty years have elapsed, threatens England's own position as a World-Power."

One has to wonder whether Mary Chestnut's influential friends, had their South triumphed and had they survived long enough to read and ponder General von Bernhardi's words, would still have considered secession a valid solution to their problems. Allegiance to King Cotton and the concept of black slavery might not have seemed so desirable if the price to be paid was to include accepting the domination of a German kaiser bent on enslaving them all.

18. The Kingdom's Displaced Persons

As for the plant whose cultivation had been responsible for the war, Southerners like Mary Chestnut's father-in-law, ruler of Mulberry Plantation and its broad acres, began the war of a secession he really did not believe in with the idea that the South should "make as much cotton as possible and send it to England as a bank to draw upon."

Under the pressure of the blockade and their need for an overseas ally, men like Chestnut were soon singing another tune. They realized that if England were to continue to receive their cotton, they would be giving it no incentive to enter the war on their side. Best that they took a lesson from the Russians who, with Napoleon's army encamped before Moscow, set their city to the torch, leaving only ashes for an enemy to seize. An artificially created cotton scarcity might help show the British where their real interests lay.

So Mary Chestnut was soon recording, "Mr. Chestnut is for a second Moscow—burning the cotton." This she had written in August, 1861. By November of the same year, she was noting, "Mr. John Raven Matthewes has burned his cotton, his gin houses, and his negro houses. Moscow complete! The Moscow idea is rampant."

By February, 1864, that Moscow idea had taken such a general hold that Jefferson Davis' political opposition within the Confederacy was imputing to him, their president, the cardinal sin of having left his cotton unburnt, apparently as some obscure sort of bribe to the despised Yankees, whom they visualized seizing the cotton and shipping it abroad for profit.

A year later, all this had become an academic question. Stripped of homes, fortunes, and slaves, thousands of once wealthy Southerners were being forced to ask themselves the sad question Mary Chestnut was putting in the last pages of her Civil War diary: "When people are without one cent, how do they live? I am about to enter the noble band of homeless, houseless refugees." King Cotton's protective hand, which had for so long sustained the South, had been withdrawn from his votaries, forcing plantation owners to face the kind of poverty other whites of the South had long had to cope with.

On the whole, Mary Chestnut's journal is a fascinating document, though a frustrating one for the reader who, having remained with it for over 800 pages, puts it down with a sense that something is lacking. What about the Southerners who did not belong to Mary Chestnut's coterie? What about those who owned no great plantations, the poor whites whose ancestors the growing, expanding plantations had crowded, bit by bit, onto the infertile sandhills? They were the dispossessed people of the South—not technically without franchise, as were the blacks, so many of whom actually enjoyed a better standard of living than the poor whites. Being jealously aware of their Southern identity, such whites were prepared to fight, even to death, in a war whose root causes—cotton and slavery—had contributed so much to their own depressed state. Technically free, technically emancipated, they threw themselves into fighting a war which, if won, could actually promise them little worth fighting for.

The people of the big plantations looked on these sandhillers tolerantly, patronizingly, sometimes compassionately, as failures, as possible cannon fodder, but never as equals. The Chestnut diary gives them the scantiest of mention. In late May, 1861, she noted of her cousin's company: "Thirty of Tom Boykin's company came home from Richmond. They went as a rifle company and were then armed with muskets. They were sandhill tackies, those fastidious ones. Not very anxious to fight with anything—or in any way, I fancy. Richmond ladies had come for them in carriages, fêted them, waved handkerchiefs to them, brought them dainties with their own hands—in faith that every Carolinian was a gentleman, and every man south of the Mason Dixon line a hero."

A few lines later, she comments on her husband's young nephew: "Johnny has gone as a private . . . He could not stay at home any longer. Mr. Chestnut was willing for him to go, because the sandhill men said, 'this was a rich man's war—and the rich men would be officers and have an easy time and the poor ones be privates.' So John C. is a gentleman private. He took his servant with him, all the same." And, of course, it was not very long before Johnny Chestnut was promoted from the ranks.

Sandhillers, having no place in Southern high society itself, had very little place in that society's thoughts and plans and would continue to have practically none as long as the planting of cotton remained the chief preoccupation of plantation owners. However, neither the Chestnuts nor any of their fellow plantation owners, either before or during the war, seemed to be aware of a new trend that was to help rebuild the South's shattered economy after the war had come to a bitter end. Nowhere in the long list of friends and acquaintances mentioned by Mary Chestnut can one find the name of William Gregg, the apostle of this new trend and a man wealthy enough to be admitted to her coterie if that alone was to be the criterion.

Of course, though, William Gregg had risen to the heights by a route which contemporary English might have deprecatingly referred to as "trade." Nevertheless, even if he did not move in the highest social circles of the Confederacy—and he probably did not concern himself with so moving—he was a loyal South Carolinian, even to the point of bringing himself to support secession, wanting the best for his state and concerned, as high society seemed not then to be, for the very sandhillers who were fiercely fighting and dying for the Confederacy.

In *Essays on Domestic Industry*, published during the 1840s first as a series in the Charleston *Courier*, then in book form in 1845, Gregg showed himself an outspoken advocate of the manufacture of cotton textiles in the South and of the poor whites who should be able to earn a decent living thereby: "It would indeed be well for us, if we were not so refined in politics—if the talent which has been for years past, and is now engaged in embittering our indolent people against their industrious neighbors to the North, had been with the same zeal engaged in promoting domestic industry and the encouragement of the mechanical arts" (Mitchell, *William Gregg*).

To the obvious question, "Where are the skilled operatives to be found?" He suggested that blacks had the capacity for acquiring the needed skills, then went on as a partisan of the poor whites:

> Shall we pass unnoticed the thousands of poor, ignorant, degraded white people among us, who, in this land of plenty, live in comparative nakedness and starvation? . . . It is only necessary to build a manufacturing village of shanties in a healthy location in any part of the State, to have crowds of these poor people

around you, seeking employment at half the compensation given the operative in the North . . . It is, perhaps, not generally known, that there are *twenty-nine thousand* white persons in the State, above the age of twelve years, who can neither read nor write— that is about one in five of the white population. (Mitchell, *William Gregg*)

During the year following the publication of his *Essays*, Gregg began trying to do something about it. He started the construction of a cotton mill in western South Carolina, not too many miles from Augusta, Georgia, where water power was available. At the time he acquired the site for his factory, there was only a single cabin in the place. But there was plenty of granite for building walls and plenty of water to be impounded behind a dam to supply power both for construction needs and for the mill machinery itself. Appropriately, Gregg named his village Graniteville.

As built in 1846, the mill was 315 feet by 55 feet, two stories high, with thick masonry walls. Starting with a single huge water wheel, about 60 feet in diameter and 12 feet across the face, it rested on an axle whose diameter measured 30 inches. Water reached the wheel through a mile-long canal, pausing on the way to power a saw mill, a grist mill, and a machine shop—the first and last contributing to the construction of both the mill buildings and the cluster of cottages which mill operatives were to inhabit. The workers were to live under the ever-watchful, kindly, paternalistic eye of the mill owner. He also built a few boarding houses such as he had noted during his visits to New England mills. Stores and churches were included in his plans.

It was not long before the factory was powered by a turbine, which replaced the first lumbering water wheel. It was lighted by gas—something many New England mills would not accept for decades—and was even heated by steam. By 1849, the Graniteville mill had been in operation for about a year. At that time, it could boast about 8,400 spindles and 300 looms, "the most improved machinery." This village built from scratch then covered about 150 acres, had two handsome churches built in the Gothic style, an academy, a hotel, ten or twelve stores and about a hundred cottages owned by the company for occupancy, at a nominal rent, by families that had members working in the mill. The original investment came to about $300,000—that is, about $32.44 per spindle.

Word about the mill and the mill village got around fast, and as soon as it was completed, families began to flock in from the country to apply for work and for homes. Of English, German, and Scotch-Irish stock, as were the plantation owners themselves, they were only a few generations removed from the hardy, independent folk who had cleared the land. Competition with Negro slave labor had cast them out and made them strangers in their own land. Now, in helping convert slave-raised cotton into usable fabrics, they were finding a way back.

That way back was not to be an easy one either for worker or mill owner, what with increasingly shrill arguments for slavery and for secession. The tragic war years loomed ahead and the equally tragic Reconstruction years followed. Nevertheless, the presence of any cotton manufacturing mill in the South was to prove a blessing to a land cut off by blockade from most kinds of imported goods. Everybody in the mill was soon overworked, office workers and operatives alike putting in fourteen-hour days. Prices paid for the goods soared, but government restrictions were choking the much-needed expansion. Even in its hour of direst need, the South was hampered by a general lack of appreciation among the ruling clique of the place industry must take in the scheme of things. Still predominantly agriculture-minded and still dedicated to the raising of cotton for distant mills whose managment was no concern of theirs, they shrilly denounced the ways of the industrial North and all but drowned out Gregg's quiet pleas.

Southerner Broadus Mitchell, historian and economist, looking back over a half century later, summed it up: "Here was a belligerent, virtually without home industry, and blockaded against foreign products, and yet enterprise in the building of new plants and even in the expanding of old ones was amazingly lacking. Much is to be attributed to the agricultural habit, something to the hope of an early end to the war, but the government policy of stifling enterprise was heavily to blame. The project of equipping and operating the finishing works at Graniteville fell to the ground when it was apparent that the profits must be transferred to others" (Mitchell, *William Gregg*, pp. 227–228).

While such restrictions imposed by a government that understood too little of manufacturing problems stifled the expansion which might and should have taken place, blockade runners were bringing in Yankee and English items, acquired in the West Indies, and selling them in the South for incredibly inflated prices. Calicoes, worth in the West Indies no more than ten or fifteen cents a yard, were sold to the South's eager buyers for nearly a hundred times as much. Even while enriching himself, the blockade runner—then a romantic figure like Robin Hood or Rhett Butler—was rewarded with high praise and was by way of becoming a national hero of the Confederacy—"while the manufacturer gives employment, comfortable homes, education, and an independent living to soldiers' wives and children, and from their labor produces an article quite as much needed, and sells it for two-fifty to three-fifty per yard, and is condemned by the public as a heartless extortioner" (Gregg, quoted by Mitchell, *William Gregg*, p. 227). Thus discouraged, the few overworked factories could not even begin to expand to meet the South's needs, and for that South a golden opportunity was lost.

Mary Chestnut had echoed the unquenchable belief of the Confederacy's ruling clique in the inevitability of England or France or both coming to the South's aid. If this aid was not to be the result of a shared aristocratic tradition, then certainly it would be the lack of cotton imports, without which their government must "topple headlong," as Senator Hammond had put it. The cherishing of such a belief and the failure to encourage any real efforts toward the self-sufficiency of the South were to prove the chief handicaps of the Confederacy.

The majority of white Southerners, unfortunately for the South, had clung to their almost feudal way of life. They paid enthusiastic homage to King Cotton and even after war came and the tide began to turn against them, continued to persuade themselves that their institutions were impregnable. Black Southerners, without rights before the law, continued to work in cotton plantations while many Northerners inveighed against the system and gave encouragement to the underground railroads by which blacks, escaped from the slavery of Southern cotton fields, were transported beyond the reach of pursuing Southern agents. The very existence of these railroads exacerbated the relationship between the people of the North and the South. They paid no heed to Gregg's timely warning that, should such a calamity as Civil War befall, "we shall find the same causes that produced it, making enemies of the nations which are at present the best customers of our agricultural productions." Trade and customers seemed not to be the concern of people who fancied themselves aristocrats.

Mitchell, looking back a half century later, summed it all up with a quote from his father, clearly one of William Gregg's disciples; "In point of view, I owe most to my Father, accepting his concise explanation that the South was overcome at Appomattox because it placed itself in opposition to the compelling forces of the age—by agency of the invention of the cotton gin held to slavery instead of liberty, insisted on States' rights in place of nationality, and chose agriculture alone rather than embracing the rising industrialism" (*The Rise of Cotton Mills in the South*, p. vii). Had the South, in accordance with the principles enunciated in the Declaration of Independence, abolished slavery, there would, he insisted, have been a happier, wealthier, as well as wiser population south of the Mason-Dixon line.

When, on April 9, 1865, after an incalculable expenditure of life and fortune, the guns were stilled at Appomattox, England and other industrial nations remained untoppled, though hardly unaffected by that war. The intervening years had been grim ones for the Lancashire mills and for the people of Lancashire and other cotton manufacturing areas. In spite of the cotton famine, even in spite of the great financial losses experienced by factory owners and manufacturers, they had not been quite able to bring themselves to declare for the South and its boasted way of life. It had become increasingly apparent that the institution of slavery was so detested by the majority of Englishmen that in the end, after much hesitation and in spite of the constant pressuring by Confederate agents in England, England could not bring itself to take a stand with the South. Thus were brilliantly vindicated William Gregg's arguments which had for so long fallen on deaf ears.

With the war ended, the cotton famine of Lancashire was beginning to approach an end and Britain could start to resume its role as chief converter of raw cotton into fabrics for the world. But it was not quite the same world, not even the same Lancashire. The long deferred and long hoped-for right to vote was at last coming to the working men of England and for this, in an indirect way, King Cotton must be given some credit.

In September, 1862, President Lincoln, having determined that "slavery must die that the nation might live," proclaimed the emancipation of slaves should take effect on January 1, 1863. Actual voting rights for the ex-slaves were to be embodied in the fifteenth amendment to the United States Constitution, which came a few years later.

As for the working men of England, they had earned the hard way a tribute paid them in 1864 by the Chancellor of the Exchequer and future Prime Minister William Gladstone: "What are the questions that fit a man for the exercise of a privilege such as the franchise? Self-command, self-control, respect to order, patience under suffering, confidence in law, regard for superiors; and when, I should like to ask, were all these great qualities exhibited in a manner more signal, even more illustrious, than in the conduct of the general body of the operatives of Lancashire under the profound affliction of the winter of 1862?"

The Lancashire workers, and with them the workers of all England as well as Gladstone's efforts in their behalf, were rewarded in 1867 with the passage of the Second Reform Bill, which gave the vote to the ordinary householder in the towns, thereby including nearly all adult male cotton workers. Some conservatives feared the consequences of so radically new a move, calling it, "a leap in the dark." Nevertheless, it was the Lancashire people's due—a people characterized by that ardent if secret Southern sympathizer, Lord Shaftesbury, as "one of the most independent people on earth; they will bear no dictation and listen to no advice unless fully assured that it comes from a sincere heart. . . . There is nothing finer on earth than a Lancashire man or a Lancashire woman" (quoted by Longmate, p. 285).

Thus the same storm that had brought emancipation to the black workers in the American South also ended with emancipation of the white cotton workers of Britain, people who were quite unlikely ever to see more of that South than the fiber which was fed to their mills.

19. The South Listens

This time, fellow Southerners were listening. They had nothing else half so promising to listen to. After all, they recognized tardily, their cotton had made fortunes for the English and for those despised Yankee manufacturers who had contributed so much toward their own defeat. Besides, their cherished old convictions that trade was not for gentlemen nor manual labor for any but blacks had left the elite of the South penniless, houseless, and homeless. Survival demanded that they change their ways of thought.

Cotton was still growing at their doors. So perhaps it was time to listen to fellow Southerners who had long been pointing out that the South should be manufacturing its own cotton into textiles. If they built factories in the midst of their cotton fields, thus avoiding most of the shipping costs for raw cotton, they might even outmanufacture and outsell the factory products of other regions.

This plan was not something to be accomplished overnight, but poverty, desperation, and the realization that henceforth the Southern people themselves must take responsibility for their future combined forces to bring the change about. It would take them twenty years after the war's end to get started, but once started they never stopped. Manufacture, once so despised, was becoming an article of faith with many Southerners.

Broadus Mitchell, in *The Rise of Cotton Mills in the South*, paints a fascinating picture of this conversion:

No undertaking was born more emphatically in the impulse to furnish work than the Salisbury Cotton Mills. All the circumstances of the founding of this factory were singularly in keeping with the philanthropic prompting. The town of Salisbury, North

Carolina, in 1887 had done nothing to recover from the war. It was full of saloons, wretched, unkempt. It happened that an evangelistic campaign was conducted; Mr. Pearson, remembered as a lean, intense Tennesseean, preached powerfully. A tabernacle was erected for the meeting, which lasted a month and, being undenominational, drew from the whole town and countryside. The evangelist declared that the great morality was to go to work, and that corruption, idleness and misery could not be dispelled until the poor people were given an opportunity to become productive. The establishment of a cotton mill would be the most Christian act his hearers could perform. (pp. 134–135)

As the *North Carolina Herald* (Nov. 9, 1887, as quoted by Mitchell) reported, Pearson, "gave Salisbury a moral dredging which made the people feel their responsibilities as they had not before, and made them do something for these folks. There had been little talk of manufacturing before Pearson came: there had been some tobacco factories in town, but they had failed. The Salisbury Cotton Mills grew out of a moral movement to help the lower classes, largely inspired by this campaign. Without the moral issue, the financial interest would have come out in the long run, but the moral consideration brought the matter to a focus." The same attitude and the same kind of response were presently taking place all over the South.

The United States Census of Manufactures for 1890 described the situation, after first noting that, although New England had been chief in the American textile industry, its spinning machinery rising from 70 percent of the total in 1860 to 81 percent in 1880, by 1890 it had dropped to 76 percent in spite of its having added two million spindles during the previous decade. So where had that other 5 percent gone? The census report went on to give the answer:

In considering the geographical distribution of the cotton manufacturing industry, the most important fact is the extraordinary rate of its growth in the South during the past decade. For a great many years, probably ever since the cultivation of the cotton plant in the South Atlantic states had a beginning, domestic spinning and weaving of coarse cotton fabrics has been a common fact in the household economy of that part of the country. Here and there small factories were established for the production of heavy fabrics. It is only in the period since the close of the civil war that mills have been erected in the South for the purpose of entering the general market of the country with their merchandise, and almost all the progress made in this direction has been effected since 1880.

The situation is the more amazing when one considers that New England had been at it for nearly a century. That first Rhode Island mill, established in 1790 by Samuel Slater under the firm of Almy and Brown, had soon been followed by others along the same stream. The first cotton mill in Fall River, Massachusetts, was built in 1813, driven by a water wheel, as were all the other mills constructed along the same stream during the following half century.

By 1902, when Thomas Young of Manchester, England, visited America to find out all he could about his hometown's American competitors, Fall River mills were boasting that they had more than one-seventh of all the cotton spindles in the Union, and that they produced two miles of cloth every minute of every working day in the year—cloth from the roughest weaves to fine dress materials. In the Fall River mills there were 105,000 work people, only 15,000 of whom were then native-born Americans; 15,000 were English, 25,000 Irish, 30,000 French Canadians, 5,000 Portuguese (undoubtedly from the families of seafaring Cape Verde islanders who had settled in New England ports). The remaining 15,000 of the total came from a variety of ethnic backgrounds. Communication of the superintendents with such a polyglot work population had already become a considerable problem.

In one mill alone, the Fall River Ironworks—so called because it had started life as an ironworks—there were over 266,000 ring spindles, an improved form of the old mule spindles of Arkwright's machines that required much less skilled labor than the mule spindles. In the same works, there were over 7,500 looms. The cloth processed in one year in their printworks, it was claimed, would encircle the globe three times with 10,000 yards to spare. Every year 47,000 bales of raw cotton were consumed there at a production rate of about twenty-three miles of thread per hour. Thomas Young noted that this was about the speed of the steamship that had brought him across the ocean.

At the time of Young's visit, the one-time whaling port of New Bedford, slightly to the south of Fall River, was constructing a large cotton mill whose electrically driven machinery was to be powered through steam turbines, the steam produced by coal-burning fur-

naces. Here each great room was to have its own motors driving the lines of shafting within that room. Without a thought for the menace in the cotton-growing and now cotton-processing South, New Bedford, weaver of fine textiles, was seeing itself as a rival to Fall River.

With electricity to power the new Northrop automatic looms, factories could now be built almost anywhere, though water power, if available to drive turbines, still remained the most inexpensive source. Young went on to visit the few mills of Maine, after stopping in northern Massachusetts to see the mills of Lowell and Lawrence and of Manchester, New Hampshire, which had so appealed to him.

Then he headed south to the states where, a scant two decades earlier, the erection of textile factories had just become an article of faith. Of this motive and of the urge to find a way back from the miseries of war and of Reconstruction, the Englishman seemed not aware. His report, though not always flattering to Southern mills, reveals how very far they had come during that brief period.

Young talked with whatever knowledgeable people he met, notably a businessman who was familiar with cotton mills of the South generally. The machinery of the Piedmont mills, he reported, "was well cared for, and the hours, as a rule, not more than sixty-six a week. But up-country both machinery and people had a hard time of it . . . He told me of children of twelve running a dozen automatic looms each for eleven or twelve hours a day, of girls of twelve drawing-in warps." Drawing-in was a highly skilled, eye-straining job by which warp threads were drawn through the eyes of the heddles that separated and guided the warp threads. In such mills, the businessman thought, the whites were, "more slaves than ever the blacks had been; they were not so well cared for, and they gained by their slavery just what the blacks used to gain—food, clothing, shelter—nothing more."

After visiting a mill in Winston-Salem, North Carolina—"small and antiquated in some ways, but very modern in others"—Young went on to a rather isolated mill near Cooleemee, owned by the tobacco tycoon, Mr. Duke. With 25,000 ring spindles and 800 looms and with cheap Southern labor overworking on Northrop automatic looms, the Cooleemee mill was producing cloth for export at a saving of 75 percent over what Northern mills charged. There seemed to be plenty of help coming in from farms of the surrounding Piedmont. As the superintendent explained to his English visitor:

> The farmers brought up large families for almost nothing; the farms produced their food; the cost of clothing in the South, where men work in a cotton shirt and a pair of cotton trousers for nine months in the year, was very much less than in the North; and fuel, when it was needed, could be had for nothing in the nearest wood . . .
>
> In this district, he said, "there are no labour laws, no school laws—in fact, no schools. But most of the mills in North Carolina, by common consent, observe a sixty-six hour week, and would rather not employ children under twelve years old."

The Englishman expressed his suspicions that this was not a humanitarian gesture but rather due to the diminishing returns inevitable if the workers were too young or forced to work too long hours. Sixty-six hours per week seemed to demand no apology. "It may be," he suggested, "that by voluntary action they are trying to avert State interference."

The recent rapid expansion of the Southern textile industry was revealed as Young continued south along the rail lines from Charlotte, North Carolina, to Columbia, South Carolina. He pointed out that within a radius of one hundred miles of Charlotte (Columbia being well inside the circle), there were nearly 300 cotton mills, containing a total of well over 3,000,000 spindles and 81,000 looms. These were about one-half of all the spindles and 60 percent of all the looms in all the Southern states in January, 1902. Within the previous six years, he was told, cotton production within that circle had doubled. It was to continue to increase, largely at the expense of Northern mills, as time would show.

"At Columbia," Young pointed out, "congregated at one end of the city, are half a dozen cotton mills, some large and new, others smaller and older, and in the newest mills may be seen at once the advantages and disadvantages attending the manufacture of cotton goods in the South. One of the advantages is the newness of the mills as compared with the mills of the North, for they have started with the most perfect machinery and the most improved methods of construction, arrangement, and so forth, whereas the older Northern mills are either old-fashioned in these respects or have been brought up to date by the expenditure of much additional capital."

Disadvantages were revealed to Young when he visited one of the show mills of the South in Charlotte. For the construction and arrangement he had much praise, but added: "Between this magnificent mill and much of the work that was being done in it there was, however, a remarkable contrast." Workers were too inefficient, too indifferent to the quality of what they were producing, too young. Blacks could find employment there if they wished to and apparently without prejudice by fellow workers. But generally they did not wish to, finding the work too demanding, the atmosphere in mills too unhealthy. They willingly left such employment to the depressed whites.

Before moving on to the Georgia mills, the Englishman made a point of visiting a smaller, more isolated mill in a place he called "Barnesville" and found conditions there much less attractive than those in the show mill near Columbia. Then he continued journeying south and west, visiting mills in Georgia and Mississippi. In New Orleans he not only studied the manufacturing there but managed to learn something of the current system of raising cotton.

"But even as far south as New Orleans," he remarked, "there are cotton mills, and the State of Louisiana has set an excellent example to its neighbours in the matter of factory legislation. A law passed in 1888 limits the number of hours during which men, women, and children may work in the cotton mills to ten a day, and prohibits altogether the employment of boys under twelve years of age or girls under fourteen years of age."

Of the planting of cotton, which had steadily been moving west since the close of the Civil War, Young learned details of the arrangements from a planter "who had been growing cotton in the Mississippi Delta ever since the war." About nine-tenths of the tenant farmers there were blacks

> although taking the South all through, fully 35 percent of the crop is raised by whites. The planter furnishes the tenant with land and seed (20 acres are the usual area for a man with a family), and also a house, fuel, water, a mule team, forage, and implements. The tenant has, besides, a garden plot given him free of charge, and enjoys the privilege of keeping fowls and pigs on the condition that they are securely enclosed. When these arrangements are made, the landlord or planter takes half the cotton crop and the tenant the other half.
>
> A mule team, I should explain, is a variable quantity. It consists of two mules for breaking and preparing the land; after that has been done it consists of one mule for cultivating the crop . . . Some tenant farmers are able to provide their own mule teams, forage and implements, and in these cases the landlord's share of the crop is only one-fourth.

Tenant farmers usually planted about four-fifths of their acreage to cotton and the remaining to maize. Apparently they rotated those two crops on the acres allotted them. Thus was many a once great cotton plantation broken up into farms of relatively small acreage to be planted, cultivated, and harvested by sharecroppers.

The English visitor from Lancashire had come to learn about cotton fiber in America—a bit as to how it was raised but mostly as to what was done to it in factories which he visited. He never mentioned that there was another source of income from the mature cotton plant. Yet by the time he was touring textile factories, cottonseed separated from the fiber by gins had become a commercial force to reckon with.

In the beginning and for many decades, that seed—two pounds to every pound of spinnable fiber—had been a nuisance or worse. A small amount, of course, was set aside to be planted for the next year's crop of fiber. The remainder piled up near the ginning houses to grow offensively rancid or to be burned, yielding an equally offensive oily smoke which, even in those pre—pollution-conscious days, was bad enough to arouse general popular protest. In desperation, some seed was dumped into the streams, which did nothing for the health of the fishes. Whatever way of disposal was tried, popular complaint followed. Something had to be done to get rid of it all inoffensively.

There was, of course, no better way than to turn the waste into profitable products. Some people had long recognized the possible value of cottonseed oil and even of the oilcake that remained after as much oil as possible had been extracted from the seed. The problem was finding an effective and inexpensive way of separating out the oil.

Back in the early 1800s, the ever-active English Society for the Encouragement of Arts, Manufactures, and Commerce had offered a gold medal to anyone who should manage to produce from cottonseed one ton of oil and five hundred pounds of oilcake, the latter presumably for cattle food. If anyone ever appeared to claim the medal, no record exists today. The offensive mounting piles of seed would be more of a stimulus than a pretty gold medal.

During the early years of the nineteenth century, half-hearted efforts were being made to start a cottonseed oil industry, but these efforts came to nothing until much later. Even in 1876, only 5 percent of the cottonseed yield was being crushed for oil. Thirty-five years later, machinery and methods had been sufficiently improved to raise that figure to 80 percent. By the 1970s, the world production of cottonseed oil had passed twenty-three million tons per annum.

Hydrogenated cottonseed oil is now a household commonplace as margarine. The crushed seed, pressed into cake and then ground, makes excellent cattle feed. And the "linters"—the fine soft fuzz attached to the seed coat—have been found to be of a cellulose so pure as to be used in surgical dressings, or, dissolved in a suitable solvent, as a base for syntheses of fibers like rayons and acetates, which were to become some of cotton fiber's own rivals.

20. Time of Reckoning

It is easy to blame the War between the States for everything that has gone wrong with the cotton industry since and, especially, for the decline of the old cotton kingdom in Georgia and the Carolinas. That decline, however, had its insidious start even before the war began. The protesting planters had only themselves to blame. They were encouraging the same repetitious agricultural methods, willfully following toward disaster the same paths that had already led, in the cultivation of tobacco, to exhaustion of the soils of Virginia and Maryland. Ignorance of the botanical facts of life might excuse those planters had they been willing to plead so ignoble an excuse. They closed their minds and went on as their ancestors had until the once great estates could no longer support either cotton or themselves.

Plants, like all living things, need not only air and water but food, which their roots must find in the soil. When nothing is left there for them to absorb, they are starved and become too stunted to compete with less fastidious weeds. Then, for all practical purposes, a crop like cotton gives up. As with people living too long under too crowded conditions, plants can sicken, become easy victims of pestilence and disease, and die off, as has happened with many a crop plant that men have transferred from the wild to plantations—cacao, coffee, quinine, or rubber, to name a few.

"There are two distinct problems in the depletion of any soil," wrote historian Avery O. Craven in 1926. "The one has to do with those factors which work immediately upon the soil to lower its yielding capacity, while the other deals with those forces which determine the use of such agricultural practices as permit destruction." In other words, if no attention is paid to maintaining soil quality to meet the demands of the plants to be grown there, disaster may be expected to follow. Where drainage is poor, plowing insufficient, or the replacement of organic matter neglected, once-fertile soils may become infertile or even toxic under continual planting and no longer able to support the same kinds of plants that once thrived there.

Craven pointed out:

> Throughout the colonial period and afterward, agriculture was based upon a single crop, produced by exploitive methods which caused yields to decline and lands to reach a condition in which planters declared them to be "exhausted." Abandonment took place on a wide scale and the planters always accepted expansion as a matter of course . . . To the evil of a single crop was added insufficient plowing and shallow cultivation, which, on loose soils and rolling lands and under heavy and concentrated rainfall, invited destructive erosion; a constant replanting of the same crop in the same soils rapidly depleted the available plant food materials and encouraged soil toxicity and the development of harmful soil organisms, and the failure to add organic matter or artificial fertilizers prevented recovery or even the checking of the work of destruction. Expansion was the only escape, and expansion from the small to the large unit and from the older to the newer regions became a normal part of life in the section; and when expansion became difficult, lowering standards of living, hardening of social lines, and conflict between the various agents in the social, economic and political life developed.

It would have been all but impossible to persuade smugly self-confident Southern planters like Senator Hammond that they themselves were in the process of destroying their own prized way of life, that they were dooming themselves, and that the pros and cons of black slavery were only incidental. In so enthusiastically paying homage to King Cotton and assuming that he must always be on his throne to reward such homage, they were actually undermining his absolute power and hastening its end.

Already in the 1840s and in Ireland, where millions depended for food upon the potatoes they could raise, the folly of planting great areas of the same crop year after year in the same fields was becoming tragically evident, though it was to take some years to understand why. It would drive some of those surviving millions across the sea to establish permanent homes there. It would permanently sour the relations between the ones who remained behind and the governing English. And all because of a microscopic organism that, finding potato plants highly acceptable to feed upon and the weather conditions of those particular years especially congenial to its growth, moved happily from plant to plant, destroying as it went.

Plant infections are nothing new. In fact, they are probably as old as the plants themselves, certainly as old as written language with which to describe them. The Old Testament tells of blightings and blastings of crops. Greek philosophers saw and described diseased plants in detail enough to be recognizable to today's plant pathologists. But neither the ancient Hebrews nor the ancient Greeks had anything better to suggest in the way of disease control than the propitiation of obviously angered gods. Even in the more modern times of the 1840s, many people were still trying to avert plagues of plants or of people through church services of prayer—because in those days no one had yet discovered what might actually be causing disastrous infections.

During those times, today's accepted idea of linking a particular collection of signs and symptoms with some particular infecting organism and with no other had yet to win sponsors in the area of human as well as of plant illness. In March, 1848, the highly literate, if occasionally misguided, editor of the English *Gardener's Chronicle*, John Lindley, expressed concern over what he considered a grave neglect of the health of gardeners:

> The larger proportion of the fevers of this country, including
> ague (malaria), and to these medical writers add rheumatism, are
> dependent upon decomposing vegetable matter. Although the
> particular form which the elements of the plants assume in order
> to produce disease has escaped detection, there can be no doubt
> of the fact . . . In the dense forests of Africa and Asia, it produces
> jungle fever. In the West Indian Islands, America, and other
> parts of the world, it produces yellow fever . . . The only condi-
> tions which are necessary to render even a small amount of vege-
> table matter capable of exerting so destructive an influence, are a
> certain amount of heat and moisture.

To do him justice, Lindley had then no way of learning the truth about the diseases he discussed nor of gauging the ignorance of those medical writers upon whom he relied. Scorn would have been heaped on anyone who might suggest that it was the mosquitoes, breeding in wet, warm compost heaps, that were the vectors of the diseases Lindley mentioned.

Investigators of plant diseases enjoy one great advantage over their fellow scientists who concern themselves with human illnesses. Plants can be experimented on without arousing horrified protests of conscience. Healthy plants could be torn apart by the dozens for study, or deliberately exposed to infection. They could eventually be sliced very thin and such sections examined under the relatively powerful miscroscopes newly becoming available.

The Reverend M. J. Berkeley owned such a miscroscope and had long been using it to study microscopic fungi, corresponding cordially with fellow French scientist Camille Montagne, who was inclined to agree that the same miscroscopic fungus that was to be seen in all the diseased potato plants they studied might be responsible for the disastrous "potato murrain." They wrote one another frequently, exchanging specimens of diseased plant tissues with a freedom that would give nightmares to anyone now concerned with the Plant Quarantine Service. Many a communication on the potato disease was to appear in the *Gardener's Chronicle* over the letters M. J. B. And soon, another correspondent, signing himself "Antifungus," was sneering at M. J. B.'s ideas. But those ideas eventually triumphed, and the science of plant pathology, or "phytopathology" was born, though not in time to keep crops as important as cotton free of disease. All the new knowledge could do was to suggest ways and means of finding out what was causing diseases and of plotting some way to control such diseases. Destroying them completely could hardly be done without destroying all the plants they infect.

One thing certain, though, was that an undernourished, overcrowded plant is an easier victim of disease than a healthy one, which has considerable power to resist infection and survive the inroads of prowling insect enemies. To raise healthy plants, good seed must be sown in good soil and the shoots be watered and cultivated properly. The more seed that is sown on given acreage, the more crowded the plants and the more easily spores of microscopic pathogens may be wafted by air currents or visiting insects from plant to plant. Larger, self-propelling enemies wait, ever ready to attack plant tissue within easy reach. Whatever may infect and possibly kill a plant in the wild affects only that single plant or, at worst, a few of its neighbors. It is in huge communities of identical plants that they become dangerously vulnerable.

Such huge communities of cotton had already brought disaster to central Georgia, which Frederick Law Olmsted described in 1861: "Decaying tenements, red old hills stripped of their native growth and virgin soil, and washed into deep gullies, with here and there patches of Bermuda grass and stunted pine shrubs, struggling for subsistence on what was once one of the richest soils in America."

A few years earlier, a slave-owning fellow Southerner was giving an equally sad description of a like situation in Alabama, a state that had become part of the Union only in 1818: "I can show you with sorrow, in the older portions of Alabama . . . the sad memorials of the artless and exhausting culture of cotton. Our small planters, after taking the cream off their lands, unable to restore them by rest, manures, or otherwise, are going further west and south in search of other virgin lands, which they may and will despoil and impoverish in like manner. Our wealthier planters, with greater means and no more skill, are buying out their poorer neighbors, extending their plantations and adding to their slave force." Already there were many farmhouses "tenantless, deserted, and dilapidated" and fields "once fertile, now unfenced, abandoned . . . Indeed, a country in its infancy, where fifty years ago, scarce a forest tree had been felled by the axe of the pioneer, is already exhibiting the painful signs of senility and decay . . . The freshness of its agricultural glory is gone; the vigour of its youth is extinct, and the spirit of desolation seems brooding over it."

Having learned nothing from the agriculture in the older states and having forgotten nothing of their unswerving allegiance to King Cotton, the planters continued to move west, expanding the old problem of soil depletion and adding new ones. Any sharecropper with little capital and a few aging acres to cultivate was doomed from the start. Only newer and more efficient methods of raising and harvesting cotton and a deeper understanding of the principles behind cultivation of that crop could save the planters, big or small, from bankruptcy.

Insect pests and such did not have to move west—they were already there, as archeologists learned when, during the 1960s, they were working in a cave situated about 150 miles south of Mexico City. On the floor of that cave, once inhabited by Zapotec Indians, they found an ancient cotton boll and within that boll an ancient weevil. Carbon-14 dating showed that both cotton and boll weevil had been alive there at about the year 900 A.D., half a millenium before any Spaniards had set foot in Mexico. What may have been the ancestry of that weevil cannot now be determined, but it was clear that cotton had long been growing in Mexico. It was equally clear that cotton was subject to the problem of the weevil, its numerous progeny seeking out other cotton bolls in the vicinity to consume and destroy.

By 1892, later generations of boll weevils had reached Brownsville, at the southernmost tip of Texas, but it was the fall of 1894 before the Entomology Division of the United States Department of Agriculture sent out an entomologist to look into the matter. The report, published in 1895, discussed the area of infestation and the life-style of boll weevils and recommended ways to discourage their further journeying into U.S. cotton fields.

The pest was truly serious, the department report said, urging the governor of Texas to start doing something about it by pushing legislation to permit quarantine and remedial work where necessary. Like politicians before and since, the governor was reluctant to admit any real threat in anything as negligibly small as a boll weevil. History might have reminded him that it was an even tinier mosquito, not a great armed enemy host, that had finally defeated the conquering Alexander the Great. In Texas, history, bent on repeating itself, simply let the boll weevil spread. By 1895, it had journeyed as far north as San Antonio and as far east as Wharton, causing almost total destruction of the crops in the infested areas and seeking out new fields to consume.

A voracious female boll weevil starts depositing her eggs in the cotton flower buds in the spring and may continue to do so throughout the growing season. Eggs hatch in three to five days, the larvae remaining inside the bolls, and feeding there for from seven to twelve days. Then comes the pupal stage, lasting from three to five days. Finally the adult weevils eat their way out of the bolls, feed for three to seven days, then mate and produce eggs to be laid in other flower buds, starting the cycle of destruction over and over again until cold weather kills the cotton plants. Weevils, however, survive, finding winter quarters in roadside trash or litter around cotton gins or near farm buildings generally.

For three-quarters of a century, the effort to control the destructive insect pests of cotton has persisted. A U.S.D.A. handbook published in 1980 claims that much has been accomplished in controlling numerous self-propelled pests of varying degrees of destructiveness—

not only boll weevils but bollworms (a recognized cotton pest since 1820); the cotton aphid, noted already in 1855 in South Carolina, Florida, Georgia, and Mississippi; the cotton leafhopper, which has been noted in Texas only during this century; the cotton leafperforator, recognized during the 1920s, and others.

Today, despite diseases and pests, cotton raising continues all across the United States, as in other areas of the world, especially below the thirty-seventh parallel of latitude. Cotton can no longer claim to be absolute monarch, the almost total subservience seen in the South having long since disappeared. Recognizing this, the people of Enterprise, Alabama, celebrated their new independence by raising a statue to the boll weevil, which had contributed so much toward freeing all Southerners, white and black, from total dependence upon the once-royal crop and prodding the South toward real independence and the age of industry.

21. The King's Rivals

Though claimed by his subjects to be undisputable king, Cotton never had the field of spinnable and weavable fibers entirely to himself. There were always rivals of a sort, some with a more ancient claim on society's loyalties, a few offering redoubtable competition, but many able to pose no serious threat to King Cotton's domains. Success in such competition would be determined not only by the special qualities of the fibers, but on the resourcefulness of people in recognizing them and their skill in finding ways to use them, trying out one kind of fiber after another.

Wool from a variety of animal pelts might have been a first choice, but animals could flee from the flaying knife whereas plants would remain where they grew. In any case, plant fibers were presently recognized and ways found to separate them from the surrounding plant tissues. In colonial America, it was the inner bark of the elm; in the South Seas, the inner bark of the paper mulberry. Stem fibers were, in temperate lands, taken from hemp or flax. In Asia or Africa, papyrus fulfilled the same need. In the Philippine Islands it was a huge, banana-like plant, there called *abacá* but elsewhere coming to be known as "Manila hemp" or simply "Manila." In any case, the variety of plants thus made use of in early times testifies more eloquently than words to the needs which drove early cultures to seek out and test the fibers of plants growing in their environment.

Already in very ancient times, people were twisting fibers into usable ropes such as those carrying cords of flax, which have been found in the remains of primitive Swiss lake dwellings and are estimated to date from about four thousand years ago. The greater structures of the East, dating probably between four and five thousand years ago—the Tower of Babel and the early pyramids of Egypt—

151

demanded, for their building, not only ramps and platforms but ropes to bind the planks into place and to help raise to increasingly great heights the massive burdens of brick and stone.

Perhaps the most intriguing use of ropes mentioned in the Bible is when Delilah, wanting to weaken Samson, wheedled out of him the secret of his strength. Somewhat cautiously he told her, "If they bind me with green withes such as have never been dried, then shall I be weak." Thus bound, however, he broke the withes "as a thread of tow is broken when it touches the fire." On a second try, he told Delilah, "If they bind me with new ropes that never were occupied, then shall I be weak." But he promptly "broke them off his arms like a thread." He should have been warned by these abortive attempts but apparently was not, and finally confessed that his strength lay in the length of his hair.

The important lesson to be learned from the disaster that followed is not so much that long hair should not be tampered with or even that men should be more reticent when chatting with women, but that in the days when Samson lived, estimated to be about eleven hundred years before Christ, ropes were being laid much as in our own day. However, the flexible plant stems, called "withes," were not yet completely outdated for bindings.

Ropes were presently playing a more explicit role in international affairs. Though wars may not have been fought primarily to gain control of lands where fiber plants grew, in contrast to spices, only guerilla wars could have been fought without the availability of ropes. Ropes bound together the boats that bridged the Hellespont for King Xerxes' invasion of Greece in 480 B.C. And ropes rigged the ships with which the Greeks defeated Xerxes at the Battle of Salamis. For many centuries thereafter, it was hemp (*Cannabis sativa*) that provided the rope fiber of choice for the navies, which would have had to remain shore-hugging and slave-powered had it not been for ropes. It was ropes that set them free to spread their sails and challenge the oceans of the world.

Hemp became so important to seafaring nations that bounties were paid to raisers of hemp, as witnessed by an act passed in 1728 by the State of Rhode Island. Thus, ironically, hemp was to grow so universally that even today drug enforcement agencies find it all but impossible to eradicate. The leaf and tender shoots of *Cannabis sativa* are, unfortunately, the source of that drug variously referred to as pot, hashish, which once was reputed to be responsible for the word *assassin*, from the Arabic *hashishin*—"hashish smokers." Presumably the assassins were men an Eastern potentate kept high on hashish so that, by controlling their access to the drug, he could send them forth to murder Crusaders and pay them for this service with hashish.

Flax, whose fibers resemble hemp's so closely that sometimes it is difficult to tell the two apart after processing, has no such sinister history. Belonging to the plant genus *Linum*, it now grows throughout the temperate regions of this earth in both blue-flowering and white-flowering varieties. The two fibers have had similar uses, coarse cloth having been woven from spun hemp and fine ropes laid of flax. Flax was one of the earliest plant fibers to become a part of woven cloth, as suggested by the still extant linen wrappings of four-thousand-year-old Egyptian mummies. Ancient Romans valued flax, and when they conquered and settled in Britain during the first centuries of our era, they tried to introduce flax culture there. Interest in flax, however, was soon to wane in most European lands until, in the eighth century, Charlemagne tried to encourage its cultivation. This time the linen industry, as well as the plant responsible for it, took root and grew apace. Flanders became the center of flax culture and of the weaving of fine linens, with the industry gradually spreading out over northern Europe. Linen was to become, during the Middle Ages, the most common of textiles, outdistancing wool, which had previously been without rival. Cotton would offer no serious competition until after the nineteenth century had dawned.

In the highlands of Mexico and Peru, where neither flax nor hemp grew native, people turned for the cords and ropes they needed to the tall-flowering, spike-leaved century plant. William Prescott, nineteenth-century historian of the conquest of Mexico, described the plant: "but the miracle of nature was the great Mexican aloe, or '*maguey*,' whose clustering pyramids of flowers, towering above their dark coronals of leaves, were seen sprinkled over many a broad acre of the table land." He went on to describe a multitude of uses, including the "thread of which coarse stuffs were made; and strong cords were drawn from its tough and twisted fibers . . . The *agave*, in short, was meat, drink, clothing, and writing materials for the Aztec."

Such was the agave of Mexico. Peru had a closely related agave. From either may come the sisal binding twine used in harvesting machines. With plenty of arid land in Peru, there was plenty of agave growing there. Garcilaso de la Vega described it, calling it "hemp" for the benefit of Spanish readers: "The plant is ugly to look at . . . The leaves are thick and about half a fathom long; they all sprout from the foot of the plant . . . The leaves might more properly be called sheaths . . . and those which are left to ripen and dry at the foot of the stem yield a very strong hemp used for making the soles of shoes, and ropes, halters and cables, and other tough products." He then describes a retting process much like that used for true flax and hemp: "The leaves that are cut before they dry are crushed and put in a running stream so as to wash away their viscosity: they then produce a different hemp which is not so coarse as the former. It serves to make slings that are worn round the head"—tump lines, of course—"and clothing whenever there is a lack of wool or cotton. It resembles the Anjou canvas which comes from Flanders, or the rougher burlap made in Spain."

It was natural for Garcilaso to compare agave fiber to cotton, for cotton fiber was a sort of standard, being in fact more spinnable than the others. What made it so was that, as the seed matures, the thousands of tubular fibers within a young boll—each fiber being a single cell anywhere from three-eighths to two and a half inches in length—dry into flat twisted ribbons that do not slip along one another when twisted together. Smooth, round fibers, notably those that grow on the tropical kapok tree and look very much like true cotton, are unspinnable. In the case of true cotton, the soft fuzz clinging to the seed coat and too fine to spin is of a pure cellulose and is used in chemical industries. For versatility, the cotton plant almost rivals the Mexican agave.

All such plant fibers have contributed toward the advance of civilizations. Without such fibers, cords could not have been made for binding or carrying, ropes could not have been formed to rig the ships and hoist the sails that were to give wings to explorers of distant seas. It was some of these navigator-explorers who were to bring home to Europe from the Far East news of a different kind of textile made from a unique fiber that was spun out to almost infinite length by a caterpillar they called "worm." Soft and shimmering like nothing Europeans had known before, fabrics woven of such material were beginning to reach Mediterranean lands perhaps two thousand years ago. Coveted by Roman women, they were sternly rejected by Roman men as too effeminate for their own apparel and too costly for their wives'. These new imported silks were then being sold for their weight in gold, so in the third century A.D. the frugal Emperor Aurelian forbade his empress to wear such costly fabrics.

For long these Romans knew so little about the fiber from which the silken textiles were woven that they believed silk grew as a fleece on trees. And why not? After all, except for wool, most of the fibers they knew and used were produced by one kind of plant or another. And in the case of silk, they had justification of a sort. Though it is not clear what special kind of tree they supposed to be producing silk, it was probably some kind of mulberry, for mulberry trees are essential to silk production. Silkworms cannot grow to maturity and spin their cocoons unless they feed on mulberry leaves. So the news that there was some important connection between tree and fiber would have immediately suggested the silken fleece theory. But when Europeans learned that a worm did it all, they were soon agog to raise their own silkworms and have their own mulberry plantations.

The Chinese, with whom the silk business had started, naturally had no intention of parting with the details of so profitable a secret as silk raising. But no secret can be kept forever, and gradually details began to leak out—first to Japan and India through the insidious subverting of a few silk workers. These workers smuggled out of China both silkworm eggs and mulberry seeds. As for the latter, though, there were already in both lands varieties of mulberry trees which might supply the essential leaves. In India especially, secrecy could not be maintained. Once the details of silk raising became known there, the recently organized Dutch and English East India companies, ever alert for profitable enterprises, would contribute their part toward spreading the good news.

By the 1770s, when the Dutch sea captain Jan Splinter Stavorinus visited India, he was able to observe silk raising—*sericulture*. He described the care and hatching of the silkworm eggs and care of the larvae after they had hatched: "As soon as they perceive a worm is about to spin"—which is shown by the worm raising its head and starting a side-to-side motion in a sort of figure eight—"they take it away from the others, and put it upon this mat . . . where it spins its ball or cocoon, which is afterwards reeled off in warm water."

It sounded alluringly simple: eggs laid by the thousands, a simple routine for keeping them and hatching them, mulberry trees growing nearby to supply the food the hatched worms needed. And when those worms reached the right stage of development, they could be counted upon to spin their silken cocoons. The only thing silk growers could not leave to nature alone was the gathering in of leaves to feed those worms, which had to be kept in enclosed containers. With this method the cocoons could be gathered in and unwound into fine filaments, which presently would be "thrown"—that is twisted into stronger threads. It all sounded like a highly profitable undertaking, particularly to Englishmen considering the wide expanses of an America that seemed extraordinarily hospitable both to trees and to worms.

King James I, who thoroughly disapproved of tobacco smoking, tried to compel the planters of Virginia to abandon tobacco raising for sericulture, which he expected to feed the looms of England. In 1623, it was decreed that any Virginia planter who failed to plant at least one mulberry tree per acre was to be fined ten pounds. By 1657, authorities were further trying to stimulate the silk-raising business by the offer of a bonus of 10,000 pounds of tobacco to any planter exporting £200 worth of raw silk or of cocoons themselves in the space of a single year; 5,000 pounds of tobacco to anyone producing 1,000 pounds of raw silk; 4,000 pounds of tobacco to anyone producing silk exclusively. Who was to raise all that tobacco and where was not stated. However, that did not much matter, for nine years later, the bounty offer was withdrawn, to be renewed in 1669, but there is no record of it ever having been claimed.

Nevertheless, silk culture did have one advantage not shared by cotton. It could be made a quite personal domestic project. No great annual plantings were necessary, to be followed by lengthy and skilled cultivation of the soil. Black slavery would never haunt the silk industry as it did the cotton. As long as there were plenty of mulberry leaves and, incidentally, plenty of children to gather them, the worms could be persuaded to spin as happily in a human living room as in a cattle shed—more so, in fact, since a more even warmth could be expected in the living room. To residents of states as far north as Vermont, where cotton could not possibly mature, raising silkworms in the home seemed a marvelous way to expand a rather limited income. It began to seem even more so when would-be silk raisers learned that there was a very special new kind of mulberry with more luscious leaves to feed to their growing colonies of worms.

The economics of silk raising might not have seemed so alluring to them had they stopped to consider the details. An ounce of eggs contained about 35,000 eggs, of which about 30,000 could be expected to hatch. Before that single ounce of worms had matured to the point of spinning cocoons, three-quarters of a ton of fresh mulberry leaves would have been gathered and fed to them, to yield from 130 to 140 pounds of cocoons. This amount, in turn, could yield no more than 12 pounds of properly reeled silk. The amazing thing is that sericulture as a private venture persisted for so long.

Yet for all colonists, ever needing new ways to make fortunes or just to keep body and soul together, silk growing had seemed the ideal way. They planted mulberry trees—what colonist couldn't?— purchased an ounce or so of eggs, then let nature take its course, supervised by children and the elderly who couldn't find other employment. Fortunes were slow in materializing. Then the beautiful dream came to an abrupt end when those wonderful new mulberry trees turned out to be susceptible to one of the plant kingdom's more serious diseases. The worms contracted diseases of their own, and the great private silk-raising craze, begun in New England in late colonial days, came to an abrupt end a century or so ago.

In more southerly colonies like Georgia and the Carolinas, where a climate more encouraging to tree growth had seemed to promise better yields, colonists had also been trying their luck at silk raising and with apparently better success. By 1735, eight pounds of raw silk were being sent from Georgia to England, there to be thrown and then woven into a textile to be presented to the queen. Presently a filature for throwing silk was built in Savannah, Georgia. Thousands of pounds of raw silk were processed before Georgia gave up in favor of the royal fiber, cotton. Of course the blame for this silk failure in Georgia can be laid at Eli Whitney's door, for with the gin he invented Georgians decided they could acquire greater profit with less personal labor than silk raising needed. South Carolina also flirted with silk raising, producing enough to be woven into material for three gowns, one of which was presented to the princess dowager of Wales. Nothing is said as to whether she ever wore it.

So silk raising as a project for Englishmen in England's colonies long remained an impossible dream, as silk itself remained a costly and coveted commodity. Its greatest value, however, was not to be in the shimmering textiles that ladies and, eventually, gentlemen wore but as an object lesson to scientists. This new seventeenth-century breed accepted nothing as impossible or beneath consideration—not even taking a lesson from a mere worm and producing artificial fibers to outrival the natural. The idea would come to Robert Hooke, member of the prestigious newly founded British Royal Society, who devoted years of his life to peering through one of those recently invented devices called microscopes, then reporting what he had seen to fellow members of that society. Hooke's *Micrographia*, published in 1665, explains in considerable detail the thoughts he had had on the subject of silk and silk-like fibers:

> Silk, seeming to be little else than a dried thread of Glew . . . a pretty kind of artificial stuff, I have seen, looking almost like transparent Parchment, Horn, or Ising-glass, and perhaps some such thing it may be made of, which being transparent and of a glutinous nature, and easily mollified by keeping in water, as I found upon trial, had imbib'd and did remain ting'd with a great variety of very vivid colours, and to the naked eye, it look'd like the substance of Silk. And I have thought that probably there might be a way found out, to make an artificial glutinous composition, much resembling, if not full as good, nay better, than that Excrement, or whatever other substance it be out of which the Silk-worm wire-draws his clew. If such a composition were found, it were certainlie an easie matter to find very quick ways of drawing it out into small wires for use. I need not mention the use of such an Invention, nor the benefit that is likely to accrue to the finder, they being sufficiently obvious. This hint, therefore, may, I hope, give some Ingenious inquisitive Person an occasion of making some trials which, if successful, I have my aim, and I suppose he will have no occasion to be displeas'd.

If Robert Hooke's fantastic idea met with no derision then, it was only because he was known as a sober and dedicated experimental scientist and a member of the socially, as well as scientifically, prestigious Royal Society. In those days, scientists did not specialize but poked into all the corners of science, perceiving many alluring paths but without the needed scientific background to follow them. Unfortunately for Hooke's dream, chemistry was hardly a science. And without a maturing chemistry, there could be no "glew" such as Hooke described. At least there could be no conscious preparation of such a material, with the preparation to be repeated at will. Certainly not even the most ingenious inquisitive person could hope to succeed in the preparation of an artificial silk at a time when chemistry, not yet fully graduated from alchemy, set most natural phenomena apart on pedestals as something mere human beings could never hope to duplicate.

Thus, until chemistry came of age two centuries later, there could be little real progress toward the planned synthesis of a natural fiber and then only effectively after scientists had humbled themselves to take lessons from the silkworm. Of course, the silkworm, if not exactly inquisitive, had always been more ingenious than any human being who would like to rival its achievements. Perhaps all nature ever intended for that worm was to produce the fiber people came to covet, but even physiologists and biochemists have not found out exactly what triggers its formation at a special time at very special spots of the worm's anatomy any more than they have learned why a cow's horns form on her head, never on her shoulders or rump. Suffice it for a textile enthusiast that a thick, gluey substance is formed in certain glands along the larva's sides and, when the worm has reached just the right age, flows to tiny openings on its head. The worm, programmed for this event, sends out two very fine threads that harden and intertwine as the head moves from side to side in the figure-eight motion.

For those who yearn to produce silky threads by their own efforts, the whys and wherefores of the silkworm's internal biological imperative may be ignored. Not to be ignored by any would-be human silkworm is the composition and molecular structure of the thread thus produced. What are that thread's component elements? How is so long a fiber built up of its component elements? Are those elements first grouped together into smaller molecules and what may such molecules be? How are they joined together?

Once these questions have been answered, there comes the sixty-four dollar question: what makes for the specially desirable qualities in the resulting huge molecule? And once that one is answered comes the climaxing one: Why should it not be possible to build up such molecules in the laboratory and to endow them with the appearance and other qualities desired? The final product should flow, as does the material within the larva's body, then harden when exposed to the air into long, strong fibers that do not disintegrate for centuries. Quite ancient silks are to be seen in many a modern museum.

The core material of natural silk is closely related chemically to the proteins in human or animal hair or to the horn destined to adorn the heads of numberless four-footed creatures. Surrounding that core, in silk, is a special albumen called sericin. Laboratory synthesis of materials like core and coating required a quite different kind of expertise from that demanded for the first artificial fibers—fibers which had the look and feel of silk but whose chemical composition was more like cotton's.

The road to these first artificial fibers began to open up only after it was discovered that cellulose, a component of cotton, could be dissolved in some solvents, then recovered either by removing the solvents (by evaporation) or by treatment with chemical reagents which could force the solution to give up the dissolved cellulose. The problem then became how to orchestrate that recovery so that the recovered cellulose would form a long fiber. It was then that human beings took a lesson from the worm. After some fibers had been prepared by dipping a needle point into such a solution, the point was drawn away while the solvent evaporated into the air from the adhering material. Later, such solutions were to be forced through fine holes in such a way that on the far side the recovered cellulose formed thin threads. It sounds obvious now, but it took chemists and textile engineers years to bring to practical reality this process which now produces those silky fibers we know as acetate, rayon, and viscose. It would take many more years to produce the kind of "glew" dreamed of centuries ago by Robert Hooke.

Actually, the first use of this method of dissolving and recovering cellulose had nothing to do with textiles. A filament which could be made to a more or less specified thickness and length seemed to be just what workers in the new applied electricity were looking for. It might be looped and sealed into glass bulbs with contacts allowed to an outside wire. Then, to the opening left in the tip of the bulb was attached a vacuum pump. Practically every molecule of oxygen was drawn out and the tube sealed off to leave a pointed tip such as a few old-timers may still recall having seen on early electric light bulbs. Once the current was turned on, the cellulose filament would char to carbon (since the oxygen needed for burning was contained within the molecules themselves). The carbon-filament lamp glowed as current passed through it, thus replacing candles and lamps.

Inventive though this process may have been as a start, its commercial future was limited, since those first filaments were produced by the needle-dipping method. A patent for the process was issued in 1855. By 1884, however, methods and machinery had been sufficiently improved for extrusion to be successful. Thus rayon, a glossy, smooth, silk-like fiber was born, the accoucheur being the French count Hilaire de Chardonnet. Interestingly, it was mulberry leaves that provided the cellulose for that first commercial venture, which began production in France in 1891.

Several decades were yet to pass before artificial fibers approximating silk in chemical composition were to come on the textile scene. Though many chemists may have been dreaming of outdoing the silkworm and though chemical industries were quite willing to invest heavily in the realization of that dream, it could not be realized until chemists came to understand, then to duplicate in part, at least, the internal structures of those huge molecules which are commonplaces in nature.

There was a time decades ago when students of biological chemistry were taught that there are three classes of solutions. One class was the "true solutions" where the dissolving material—salt, say, or sugar—seems to disappear altogether in the solvent. At the other end of the scale was a mixture containing a suspension of fine particles that would, given enough time, settle out, as sand in water. The rate of settling depended on the size of the suspended particles, the smallest taking a very long time to settle. Between these two extremes were "colloidal suspensions," which remained cloudy and never settled out. Here the particles were presumed to be larger than the molecules in true solutions, yet not of a size to be measured under any visual microscope. Presently, it was suggested that these suspended colloidal particles might just be huge molecules of a size not then to be estimated and whose very existence was at first derided.

As organic chemistry progressed, however, the existence of those huge "macromolecules" was accepted and their chemistry began to take the center of the stage. It was here that some of modern chemistry's most exciting advances, both theoretical and practical, were to be made. This was the new polymer chemistry—new not because polymers had never before been prepared in the laboratory but because it took a very ingenious and inquisitive chemist to recognize them for what they might be and to see a link between internal structure and external behavior.

Previously, during the many decades when organic chemistry was growing from practically nothing, a student who attempted to prepare certain known chemical compounds might end up with a sticky, gooey mess that clung to the insides of the reaction vessel and could be removed, if at all, only with much difficulty and great patience. Of course the student threw out the mess, probably cursed, then started all over again. Today, in view of later advances, the student might cherish the mess, study its properties, and ask whether it might not be worthwhile to prepare more of the same for study.

Chemists gradually accepted the fact that there need be no limit to the possible size of molecules they could build up and that properly selected small molecules of well-known composition might somehow be persuaded to join together in some foreordained pattern to form great long-chain molecules such as nature easily produces in various forms. Once this idea was accepted, some chemist was bound to take the next step of preparing macromolecules to order. It was no simple task, for the starting materials, reaction temperature and pressure, timing, and other conditions all had to be just right for the reaction to go as desired and end as desired.

It was not an insoluble problem for the brilliant young Wallace Carothers, who in the early 1930s intentionally produced a gooey mess, then tested its qualities by forcing some of it through a hypodermic syringe. This process formed a lustrous, elastic, strong, silk-like fiber that, on cooling, could be further strengthened by stretching. It was determined that stretching served to align the molecules along the length of the thread. This was nylon—further labeled "66" because it was built up of two types of small six-carbon organic molecules.

It was, of course, a big and costly jump from the hypodermic produced thread to the factory-planned nylon we know, but it has since repaid the investment many times over. Furthermore, once the trick had been mastered, a host of other polymeric fibers were to follow, each one tailored to some special needs. Most of these have been the result of high-temperature polymerization of smaller compounds that occur frequently in nature. When this has taken place, the mass is cooled, cut into chips, melted, forced through spinnerettes or formed into sheets. Either type is then stretched to give the needed strength through aligning the molecules.

Dacron, orlon, nylon, saran, and spandex are all synthetic polymers of the class to which natural silk belongs. Silk started it all, without raising the temperature or changing the atmospheric pressure, by mulberry-feeding worms whose efforts it has taken human beings centuries to copy, though not quite to duplicate.

Like other fibers, these synthetics may have characteristics that some people find less than desirable. But they have in a large measure been modified by blending with one or another natural fiber, producing a final fabric with properties closer to the heart's desire.

Cotton plays a large role here. But, on the whole, that domineering plant which was once acknowledged as king has had to take a secondary role to synthetics.

We should not, though, overlook the fact that in the production of synthetic fibers a great deal of energy must be expended. Energy now being in ever-increasing demand and diminishing supply, it could just be that, in textiles as well as in foodstuffs, people will be swarming back to the natural products. Already expensive pure cotton fabrics are to be met with, and clothing made from this fabric brings exorbitant sums.

Who knows but that a modern cotton kingdom may presently come to replace the old?

Bibliography

Author's Note: Most of the books consulted in the course of this work are listed here. Two (Radcliffe and Thackrah) were not seen as published but quoted from Baines. Not listed are various encyclopedia articles, notably those on synthetics, which, being covered by patents, are not generally discussed in detail. Excerpts from books here listed with their Spanish titles are given in my own translation.

Adams, Charles Francis. *Trans-Atlantic Historical Solidarity*. Oxford: Clarendon Press, 1913.

Agriculture Handbook 515. U.S. Department of Agriculture Annual Conference on Cotton Insect Research and Control. 1947–1974.

Aldersey, Laurence. "The Second Voyage of Master Laurence Aldersey to the Cities of Alexandria and Cairo in Aegypt, Anno 1586." In Hakluyt, *Principall Navigations*, pp. 224–227.

Arnold, R. Arthur. *A History of the Cotton Famine from the Fall of Sumter to the Passing of the Public Works Act*. London: Saunders, Otley & Co., 1865.

Baines, Edward. *History of the Cotton Manufacture in Great Britain*. London: H. Fisher, R. Fisher & P. Johnson, [ca. 1845].

Bates, Edward C. "The Story of the Cotton Gin." *New England Magazine*, n.s., vol. 2 (May 1890), pp. 288–292.

Benavente, Fray Toribio de. *See* Motolinia.

Bernhardi, Friedrich Adam Julius von. *Germany and the Next War*. Translated by Allen H. Powles. New York: Longmans Green & Co., 1914.

Bright, John. *The American Question*. Boston: Little Brown, 1865.

———. *Speeches on questions of Public Policy by the right honourable John Bright M.P.* Edited by J. E. T. Rogers. London: Macmillan & Co., 1869.

Brown, Henry Bates, and J. O. Ware. *Cotton Growing*. New York: McGraw-Hill Co., 1958.

Bruchey, Stuart. *Cotton and the Growth of the American Economy*. Forces in

Economic Growth series, edited by A. D. Chandler, Jr. New York: Harcourt, Brace & World, Inc., 1967.

Byram, H. K. *An Essay on the Culture and Manufacture of Silk*. U.S. 30th Cong., Exec. Doc. No. 54, Appendix No. 10. Washington, D.C., 1848.

Carlyle, Thomas. *Critical and Miscellaneous Essays*, vol. 4, *Chartism*. New York: Charles Scribner's Sons, 1904.

Chestnut, Mary Boykin. *Mary Chestnut's Civil War*. Edited by C. Vann Woodward. New Haven: Yale University Press, 1981. (Originally published in 1905 as *A Diary from Dixie*.)

Chilton, John. "A Notable Discourse of Master John Chilton Touching . . . Memorable Things of the West Indies." In Hakluyt, *Principall Navigations*, pp. 587–594.

Cotton, Edward. "Certayne Remembrances of an Intended Voyage to Brasill and the River of Plate . . . 1583." In Hakluyt, *Principall Navigations*, pp. 187–188.

Craven, Avery C. *Soil Exhaustion as a Factor in the Agricultural History of Virginia and Maryland*. Urbana: University of Illinois Press, 1926.

Darwin, Erasmus. *The Botanic Garden* . . . 2 vols. Edited by D. H. Reiman. New York: Garland Pub., Inc., 1978. (Facsimile of 1870 edition.)

Defoe, Daniel. *A Review of the State of the British Nation*, vol. 4. New York: AMS Press, 1965. (Facsimile of 1708 edition.)

Desaguliers, John Theophilus. *Course of Experimental Philosophy*. 3d ed. 2 vols. London: A. Millar, 1762.

Diaz del Castillo, Bernal. *Verdadera relación del descubrimiento y conquista de la Nueva España y Guatemala*. 2 vols. Biblioteca "Goathemala." Guatemala City, 1933–1934. (Originally published in Madrid, 1632.)

Dickens, Charles. *American Notes*. Philadelphia: J. B. Lippincott Co., 1891.

Dircks, Henry. *The Life, Times and Scientific Labours of the Second Marquis of Worcester*. London: Bernard Quaritch, 1865.

Eisler, Benita, ed. *The Lowell Offering: Writings of New England Mill Women (1840–1845)*. Philadelphia: J. B. Lippincott Co., 1977.

Fletcher, Francis. *The World Encompassed by Sir Francis Drake* . . . London: Nicholas Bourne, 1628. Reprint, Facsimile Bibliotheca Americana, Cleveland: World Pub. Co., 1966.

Garcilaso de la Vega. *Royal Commentaries of the Incas*. 2 vols. Translated and edited by Harold V. Livermore. Austin: University of Texas Press, 1966.

Gerard, John. *The Herball or Generall Historie of Plantes . . . enlarged and amended by Thomas Johnson, apothecarye of London, 1633*. Facsimile. New York: Dover Publ., 1975.

Gladstone, William Ewart. On Borough Franchise Bill. *Hansard's Parliamentary Debates*, ser. III, vol. 175, p. 326. London, 1864.

Hakluyt, Richard. *Principall Navigations, Voiages, and Discoveries of the English Nation*. London, 1589. Facsimile with introduction and index, edited by D. B. Quinn and R. A. Skelton, 2 vols., Cambridge: Cambridge University Press for the Hakluyt Society, 1965.

Hammond, James Henry. *Selections from the Letters and Speeches of the Honorable James Henry Hammond*. New York: John F. Trow, 1866. Reprint, Spartanburg, S.C., 1978.

Hareven, Tamara K., and Randolph Langenbach. *Amoskeag: Life and Work in an American Factory City*. New York: Pantheon Books, 1978.

Herodotus of Halicarnassus. *History* . . . Translated by George Rawlinson. Edited by Manuel Komroff. New York: Tudor Pub. Co., 1934.

Hooke, Robert. *Micrographia* . . . London: Jo. Martyn & Jo. Allestry, 1665. Facsimile reprint, Early Science in Oxford series, vol. 13, 1938.

Knight, Stephen A. "Reminiscences of Seventy-one Years in the Cotton Spinning Industry." *Scientific American*, vol. 61 (1906), pp. 26ff.

Larcom, Lucy. *A New England Girlhood*. Boston: Houghton Mifflin, 1889.

Longmate, Norman. *The Hungry Mills: The Story of the Lancashire Cotton Famine, 1861–1865*. London: Temple Smith, 1978.

Lowell Offering, Mind among the Spindles: A Selection. London: Charles Knight, 1844.

Martineau, Harriet. *Society in America*. London: Saunders, Otley & Co., 1837.

Miles, Henry A. *Lowell As It Was and As It Is*. Lowell, Mass.: Powers & Bagley, 1845.

Mitchell, Broadus. *The Rise of Cotton Mills in the South*. Baltimore: Johns Hopkins Press, 1921.

———. *William Gregg, Factory Master of the Old South*. University of North Carolina Press, 1928. Reprint, New York: Octagon Books, 1966.

———, and George Sinclair Mitchell. *The Industrial Revolution in the South*. Baltimore: Johns Hopkins Press, 1930. Reprint, New York: Greenwood Press, 1968.

Motley, John Lothrop. *History of the United Netherlands*. 4 vols. New York: Harper Brothers, 1867.

Motolinia (Fray Toribio de Benavente). *Historia de los indios de la Nueva España*. Mexico City: Editorial Chávez Hayhoe, 1941.

Olmsted, Frederick Law. *The Cotton Kingdom: A Traveller's Observations on Cotton and Slavery in the American Slave States*. New York: Mason Brothers, 1861. Reprint, New York: Alfred Knopf, 1953.

Owen, Edgar R. J. *Cotton and the Egyptian Economy, 1820–1914*. London: Oxford University Press, 1969.

Pearse, Arno S. *Cotton in North Brazil* . . . Manchester, England: Taylor, Garnett, Evans & Co., n.d.

Pliny. *Natural History*, vol. 5. Translated and edited by W. H. S. Jones, 1956.

Prescott, William H. *History of the Conquest of Mexico.* 3 vols. 1843. Reprint, Philadelphia: J. B. Lippincott Co., 1895.

Radcliffe, William. *Origin of . . . Power Loom Weaving.* Stockport, 1828.

Robinson, Harriet H. *Loom and Spindle: Or, Life among the Early Mill Girls.* New York: Thomas Y. Crowell, 1898.

Silver, Arthur W. *Manchester Men and Indian Cotton, 1847–1872.* Manchester, England: Manchester University Press, 1966.

Stavorinus, Jan Splinter. *Voyages to the East Indies, 1768–1771.* 3 vols. Translated and edited by Samuel H. Wilcocke. London, 1798. Reprint, London: Dawsons of Pall Mall, 1969.

Stephen, Leslie. *Life and Letters of Leslie Stephen.* Edited by William F. Maitland. New York: G. P. Putnam's Sons, 1906.

Thackrah, Charles Turner. *The Effects of Arts, Trades . . . and of the Civic States and Habit of Living on Health and Longevity.* London: Longmans & Co., 1832.

Thane, Elswyth. *Nathanael Greene: The Fighting Quaker.* New York: Hawthorne Books, 1972.

Thayer, William S. *Rep't in Letter of Sec'y of State on Commercial Relations of U.S. with Foreign Countries for the Year Ending Sept. 30, 1862.* U.S. 37th Cong., 3d Sess., Exec. Doc. No. 63, 1862–1863.

Trollope, Anthony. *North America.* New York: Harper Brothers, 1862.

U.S. 22nd Cong., 1st Sess. Committee Rep't No. 481, Dec. 1831. *The Arguments of the South.*

U.S. 27th Cong., 2d Sess. Rep't No. 461, Committee on Manufactures, 1841–1842.

U.S. 30th Cong., 1st Sess. House Exec. Doc. 54. *Ann. Report of the Commissioner of Patents for the Year 1847.* Washington, D.C., 1848.

U.S. 34th Cong., 1st Sess. House Rep't No. 182. *Alleged Assault upon Senator Sumner.* Washington, D.C., 1856.

U.S. 52nd Cong., 1st Sess. House Misc. Doc. No. 340, part 22, 1891–1892. *U.S. Census of Manufactures, 1890.*

Ware, Caroline F. *The Early New England Cotton Manufacture.* Boston: Houghton Mifflin, 1931.

White, George Savage. *Memoir of Samuel Slater, The Father of American Manufactures.* Philadelphia, 1836. Reprint, New York: Augustus M. Kelley, 1967.

Wilkinson, Frederick. *The Story of the Cotton Plant.* New York: D. Appleton, 1903.

Young, Thomas M. *The American Cotton Industry: A Study of Work and Workers Contributed to the Manchester Guardian.* New York: Charles Scribner's Sons, 1903.

Index

DATE DUE